A GENDER HANDBOOK FOR WESTERN MAN

Elder George

MAI Publishing Divsion

The opinions expressed in this manuscript are those of the author and do not represent the thoughts or opinions of the publisher. The author warrants and represents that he has the legal right to publish or owns all material in this book.

A GENDER HANDBOOK FOR WESTERN MAN

2nd printing

MAI Publishing Division

maipublishing@aol.com

ISBN-13: 978-1-61286-176-0

Printed in the United States

To My Sons and Daughters

May the knowledge contained herein motivate you to create a more natural way of life for humankind.

TABLE OF CONTENTS

FOREWORD

This handbook presents a specific truth or bit of knowledge called the principle of gender and the patriarchal structure that emanates from it. This knowledge is directed at Western thought, because Western thought exhibits an almost total lack of understanding of the purpose of human existence and the integral part that gender plays in fulfilling that purpose.

Like all knowledge, the principle of gender is unseen and does not lend itself to "proof" by material means. Knowledge can only be understood. The great American poet Ralph Waldo Emerson said, "I trust that I shall never utter a statement of truth that will need to be proved."

Examples and parables help to make one aware of the truth; they massage the psyche causing it to open up to the truth within. None of the great spiritual texts contain provable data, but they contain many verbal illustrations of the truth that they present.

Western thought on the other hand focuses on information, which consists of compiling data about material things such as how to bake an apple pie or build a space ship. Due to the increase of communications vehicles, the compiling of data has become very rapid and our present era is known as "The Information Age;" it is never referred to as the truth or knowledge age. It possesses very little knowledge or truth because it looks for it in the wrong place, as will be described in this handbook.

In accordance with the forgoing, I presented that portion of the truth that has been revealed to me as clearly as I could and then explained it. To give illustrations of the manifestation of this truth I have included 39 essays located at the end of the book that I wrote over the last six years, which appeared as columns in newspapers or as blogs on the internet. My original plan was to cross-reference these essays with the chapters that they most apply to; however, I soon realized it to be a highly subjective task and decided to let the reader make his or her own decision as to which chapters the essays most supported.

Accept what sounds true to you of what your read herein and reject or preferably set aside for another time that which you cannot now accept. Take what feels comfortable to your understanding and live accordingly. You will begin to see the world differently and by changing the way you live you will influence society to start functioning in the natural gender based structure known as patriarchy.

I hope that your journey through this book will be enjoyable, interesting, enlightening, and motivating.

Elder George 3-23-10

PREFACE

Western society is imploding and does not know why.

Aberrant behavior has become the norm and statistics that would normally produce shock among the populace now only receive casual acknowledgement and general acceptance.

The United States has the highest incarceration rate in the world; its prisons contain 2.4 million men. One person in six between the age of 18 and 30 has a relationship with the law. America has 5% of the world's population and 25% of its prisoners; it also has the most violent boys in the world.

Western thought has produced utter defilement of the female beginning in grade school where 10 year-old girls participate in sex clubs; continuing to adulthood where streams of women journey to California, the world's Mecca of pornography, to serve as interns in X rated films in order to build up resumes necessary to obtain work in big city adult nite clubs; to the unprecedented rate of unwed motherhood, which spawns juvenile problems and creates stress induced depression upon their mothers. Upwards of 10 million women—about 10% of the adult female population—suffer from debilitating depression in the United States; in Europe that malady afflicts 15% of the adult female population. Three million American girls now suffer from depression. The core group of depressed women comes from the white, single, middle-class. Western society has the distinction of containing the most miserable, unhappy, and confused women in the world.

According to the World Health Organization mental illness has become the number one health issue in England, the United States, and Canada. The French take anti-depressants at a rate two and a half times that of the English, and the unwed motherhood rate in France has reached 50%. Three European Union nations -- France, Spain and Portugal -- do not prosecute consenting adults for incest, and Romania is considering following suit. It will soon become the norm in the Western world to conduct sexual relations between mother and son, father and daughter and brother and sister. Livestock raisers know that this type of inbreeding weakens the herd, but Western society seems oblivious to any negative consequences.

Air, water, and food pollution and the accompanying respiratory and chronic illnesses they produce are accepted as part of the cost of a thriving economy. A society whose inhabitants are willing to suffer physical disability and die younger because it results in greater material accumulation would be called morally deficient and mentally ill. The Western world is that society.

To all who read this preface, regardless of your age, whatever you considered to be wrong with society at the date of your birth, has gotten worse. Whatever personal standard you have in judging society you will find that conditions have deteriorated. The prison population, numbers of unwed mothers, divorce, adultery, and substance abuse have all increased. Academic performance, marriage rates, physical and emotional well-being have all decreased. You live in a society incapable of correcting any problem; a condition that warrants serious reflection but does not even get acknowledged.

Ethics have disappeared in Western society including its religious institutions. Divorce among Christians is higher than among those who have no religious affiliation whatsoever. The divorce rate among the ten states that comprise the Bible Belt is 50% higher than the national average. Divorce among protestant clergy is on a par with the laity. The incidence of child abuse by Catholic clergy continues unabated. On the day that I write this preface Dublin Archbishop Diarmuid Martin slammed Irish Catholic orders for concealing their culpability in decades of child abuse. Ten percent of the American population now consists of ex-Catholics. American Jews divorce at a rate

10% higher than Christians, and American Muslims have seen a significant increase in their divorce rates.

The government exhibits a lack of ethics from the lowest to the highest levels. The Mayor of New York distributes 2.5 million condoms to school children and the President uses taxpayer money to bailout Wall Street financial institutions whose problems were created by greed driven fraud. Corruption in state and local government has become a way of life. Ethics cannot be found anywhere.

Men who devote themselves to the propagation and preservation of the species establish ethics. There is no place else for ethics to come from. Western thought—which worships the material world and developed a genderless society—is devoid of ethics. It has no sense of right and wrong. It attempts to make up for its ignorance of matters spiritual by developing laws for its societal conduct. Legal and illegal have replaced right and wrong leaving society without any meaning on which to base its behavioral conduct. Solomon said a nation without vision shall perish. Western society is that nation and is perishing.

Western thought has been singled out in this handbook because it clearly demonstrates that it has no concept of gender or matters unseen. The entire scope of Western thought fits within the narrow confines of a finite material world. After I wrote my first book *Dear Brothers and Sisters: Gender and Its Responsibility*, I came to the realization that Europe represented the feminine principle, which explained the difficulty people of Western extraction had in understanding patriarchy; whereas, the audience most understanding of gender consisted of non-Western immigrants and American born blacks—people who had a cultural history of extended family and the patriarchal structure necessary to maintain it. This receptivity to my message was born out by the fact that I wrote regular weekly columns for seven ethnic newspapers.

Western society has no concept of the extended family or tribal way of life. Anywhere that extended family exists in Europe it owes its origin to either an African or Asian influence. In Sicily and southern Italy the extended family tradition is a remnant of the Carthaginian occupation. In the Balkan countries the extended family tradition stems from the occupation of the Ottoman Turks.

Family oriented Europeans extol the nuclear family, an aberration of nature that has never worked; it cannot sustain itself and requires outside support. Western society attempts to provide this support by the development of institutions and outside family care such as orphanages, after-school programs, counseling, homes for unwed mothers, old age homes, and a variety of prisons. Europe has exported the institutionalized care of human beings. India did not have old age homes before the British arrived nor did Africans have orphanages before the Europeans arrived; they had no prisons either.

Western society has become completely institutionalized and its government has become directly or indirectly involved in the control of these institutions; a condition brought about by the gross ignorance of gender and the patriarchal way of life necessary to properly sustain and develop humankind.

At this point I will pause to acknowledge that much of this preface has focused on negative conditions, and intentionally so. These conditions have become normal even though they are not natural; once a sufficient proportion of society accepts the unnatural as normal it loses any incentive to change. America and the Western world have accepted much that is unnatural, as the ensuing paragraphs will illustrate.

The following is an excerpt from the 1971 edition of the Encyclopedia Britannica: "In the United States, where crime rates have been among the highest observed in modern societies, there are well over 200,000 prisoners confined in more than 200 state and federal prisons and reformatories, with an annual increase in the prisoner population of approximately 4,500. Current trends indicate that the number of prisoners will continue to increase."

The number of prisoners did increase and by 1976 total incarceration reached 250,000. The prison population doubled every eight years thereafter until 2000 when it reached two million. The rate of increase declined for a while but has started to climb again and the prison population in 2009 stands at two and a half million. Only a conquering army would imprison such a proportion of men and yet there is hardly an outcry from politicians, the media, or the public. We have come to accept this high prison population as normal; hence we make no effort to change it.

The unwed motherhood rate in my youth was one and a half percent. By the early sixties it had doubled to three percent and by 2000 it had reached 30%. It now stands at 40% and few people complain about it; unwed motherhood has become the norm. Very little effort is being expended to change these conditions.

Unwed mothers do not live in an environment conducive to nurturing children. They turn the "care" of these children over to the city, state, and nation in various after-school, preschool and in-school programs. These children no longer receive love; they act accordingly. America now has the most violent boys in the world, three million girls suffer from depression and one million school children receive the drug Ritalin on a regular basis. These conditions have been accepted and the trend towards institutionalized care of children is increasing.

High incarceration rates, high levels of unwed motherhood, and children who do not receive nurturing are aberrations of nature. These conditions have been accepted. They are normal to our society. Unless these aberrations and the multitudes of others that plague our society are addressed and attacked, society will show no resolve to make changes.

I will present one more example—a personal one—of the acceptance of unnatural as normal. American Medical Association figures indicate that the average person over the age of 65 takes six to eight prescription drugs. As of the writing

I am 80 years old and do no take any drugs; therefore, I am an anomaly. The norm is to take drugs. This normalization of the aberration of taking drugs has led to a financial healthcare crises part of which is the cost of medicinal drugs. Because taking medicinal drugs is considered to be normal the issue addressed by society and its government has become how to pay for the cost of drugs rather than how to lower their use.

The conduct of Western society has become an aberration of nature. It chooses to call this aberration "normal."

In order to revert to a more natural way of life Western thought must change; thinking used to produce an issue cannot be used to resolve that issue. That is the meaning of Jesus' reference to being reborn. Western thought has created the problems its society faces; only a change in thought will eliminate those problems. It must come to realize that what it considers "normal" behavior is an aberration of natural behavior

The objective of this handbook is to provide the basis for a change in thought necessary to resolve the negative issues of the Western world. It will define, explain, and then illustrate the principle of gender in action and the natural patriarchal way of life necessary to propagate and preserve the race.

Chapter I

10 MAXIMS OF GENDER FUNDAMENTALS

The purpose of human existence is to grow spiritually. In our present state it cannot be done outside of our physical bodies; therefore, the primary human activity is the propagation and preservation of the species, which is also the primary activity of the universe. The universal principle of gender and the patriarchal structure that emanates from it makes possible this universal activity.

Ten maxims of gender fundamentals:

1. Gender came into existence upon the creation of the physical universe; it enables all movement, expansion, and propagation.
2. For any activity to occur an assertive influence must act upon a receptive entity. Two assertive influences cannot produce anything, nor can two receptive entities.
3. In the consideration of relativities such as hot and cold, light and dark, and noise and quiet there can only be one control influence in each set.
4. The masculine gender represents the assertive influence of the universe and the feminine gender represents the receptive entity. Patriarchy is the natural interplay between the two genders.
5. Any area of activity has only one assertive influence but can have many receptive entities.
6. Everything produced in the universe results from a pregnancy. The masculine gender initiates the pregnancy and the feminine gender brings it to fruition and nurtures it.
7. The receptive feminine gender comprises all that can be seen. The direction of its movement depends upon the

unseen influence of the masculine gender. The nature of its movement depends upon its inner qualities.

8. The feminine gender materializes and personalizes the impersonal concepts of the masculine gender.

9. The "I" of the personality reflects the masculine gender and the "ME" reflects the feminine gender.

10. The masculine and feminine genders have characteristics that serve as attributes enabling them to fulfill their respective functions of propagating and preserving the species on its infinite journey of spiritual growth.

Chapter II

EXPLANATION OF MAXIMS

MAXIM 1

Gender came into existence upon the creation of the physical universe; it enables all movement, expansion, and propagation.

The universe pulsates with energy and movement as everything in it from the smallest atom to the largest nebulae continually expands and propagates. Life exists everywhere and in everything; nebulae spin off new suns, suns spin off new planets, and planets develop new life. A motivating energy operates behind this active and expanding universe. Nothing just happens; every effect results from a cause.

The tendency to look at the movement of the universe as "nature" or something that just happens inhibits our understanding of all existence. Nothing just happens. Life would be extremely precarious if movement "just happened." It would also imply an absence of intelligence in universal activity, instead of an understanding that intelligence pervades all activity.

The universal principle of gender provides the basis for the functioning of the physical universe.

MAXIM 2

For any movement to occur an assertive influence must act upon a receptive entity. Two assertive influences cannot produce anything, nor can two receptive entities.

The term gender comes from the Latin root meaning to produce; in order for production or movement to take place an assertive influence must act upon a receptive entity. An example would be the movement of an object, say a cup sitting upon a table. The cup sits upon the table in a perfectly receptive state and has no will. Unless some assertive influence comes in contact with that cup it will sit upon the table forever. Adding another receptive entity such as a second cup will not produce any movement. Only the application of force by an assertive influence can move the cups. If there were two assertive influences and no cups, no movement would result because there would not be anything to move.

In order for anything to happen in the universe an assertive influence must act upon a receptive entity. That receptive entity in one way or another provides for the propagation and nurturing of the race.

An observation of movement of any sort will find that an assertive influence acts upon a receptive entity. Getting into a car and driving it illustrates an assertive influence (the driver), acting upon a receptive entity (the car). The car is completely receptive; it can be moved forward or backward or turned to the left or right. If the car were not receptive, the driver would not have the ability to go where he wants. The union of the assertive driver and receptive car produced a vehicle in motion that will bring the driver to his chosen destination.

At first glance it might appear that only objects are receptive entities; but observation will show that all life can fit into this category depending on its relationship to an assertive influence. The Sun acts as the assertive influence in regards to the Earth; it holds it in orbit, moves it through space, and provides the energy for it to function. Soldiers act as receptive entities to their commanders. The governed act as receptive entities to those who govern. Females react as receptive entities in relation to males.

MAXIM 3

In the consideration of relativities such as hot and cold, light and dark, and noise and quiet there can only be one control influence in each set.

There is no such entity as cold. Cold is the absence of heat. We either add heat to something or remove the heat but we can't make cold. Refrigerators remove heat from the food stored in them and eject it into the kitchen. Air conditioners remove heat from the room and eject it to the outside. The absence of heat results in lower temperatures, which we refer to as colder. We cannot make dark, we can only add or remove light. We cannot make quiet, we can only add or remove sound.

Two opposing entities cannot operate in one area of activity. There can only be one assertive influence in any field of activity. If cold were an entity then it would be impossible to set a steady temperature. If dark were an entity then it would be impossible to set a steady level of light. If quiet were an entity then it would be impossible set a steady level of sound. A stable situation can only have one control factor.

There can only be one god in the universe, one Sun in the solar system, one chief of a tribe and one head of a family. Anything other than one results in chaos and a weakening and dissolution of the structure.

MAXIM 4

The masculine gender represents the assertive influence of the universe and the feminine gender represents the receptive entity. Patriarchy is the natural interplay between the two genders.

The masculine gender manifested in the male in all species initiates movement, develops structure, maintains order, and defends the community. Occasional exceptions might appear to occur, but on closer examination it will be found that the male leads in all species.

Whether family, flock, herd, pride, school, gam, gaggle, or troop, a male always leads it. The function of the male is to provide the environment and means for the female to bring forth life and nurture it. The attributes inherent in the male provide the authority that enables him to take the action necessary to fulfill his responsibilities. This activity is called patriarchy and represents the natural relation ship of gender in the universe.

MAXIM 5

Any area of activity has only one assertive influence but can have many receptive entities.

A Solar system can have many planets, but only one Sun. A nebula has many Suns but they all conform to its assertive influence. The universe contains many nebulae, but they all conform to the greater power that motivates and guides them. In the atom the many electrons rotate about one nucleus.

The structure of the universe is patriarchal and for the most part polygamous.

There is no such entity as an independent feminine anything in the universe. There are Suns without planets, but no planets without Suns. Every electron must have a nucleus to rotate around. The feminine receptive nature will naturally be attracted to the assertive masculine principle. It needs the masculine principle to provide the environment and means for it to function.

The function of the feminine gender is to bring life into this world and nurture it.

MAXIM 6

Everything produced in the universe results from a pregnancy. The masculine gender initiates the pregnancy and the feminine gender brings it to fruition and nurtures it.

An example of gender in action in the natural universe occurs when the masculine assertive North Pole sends a charge to the feminine receptive South Pole, which gets pregnant and produces magnetism. An example of gender in action in man-made products occurs when the masculine assertive anode in a storage battery sends a charge to the feminine receptive cathode, which gets pregnant and produces electrons. Bees pollinating flowers exemplify the receptive feminine gender becoming impregnated by the masculine gender. All movement, production, and expansion result from the interplay of gender.

MAXIM 7

The receptive feminine gender comprises all that can be seen. The direction of its movement depends upon the unseen assertive influence of the masculine gender. The nature of it movement depends upon its inner qualities.

Mother earth utilizes its productive ability to nurture the race. The bounty of mother earth can be touched, tasted, smelled, heard, and seen. The Sun holds the earth in orbit, moves it through space, and provides the energy for it to produce.

The receptive feminine gender reflects the unseen assertive masculine influence. The entire physical universe represents the feminine gender in action, which is a manifestation of the thoughts, energy and direction of the unseen masculine influence as imparted into its self-expression..

The removal of the masculine gender results in the dissolution of that which can be seen.

MAXIM 8

The feminine gender materializes and personalizes the impersonal concepts of the masculine gender.

Thought, when verbalized or written becomes feminized; it becomes limited by the vehicle of material expression. The unseen concept becomes lost in the attempt to explain it in material terms.

All religions are feminine. Infinite truth cannot be expressed in finite terms. Truth when materialized becomes information. Information can be recited and proven. Truth cannot be proven; it is understood. The relationship of religion to spiritual texts and words of the prophets represents the relationship of the seen to the unseen. One emanates from the other, but is not the other.

Religious dogma represents a materialistic interpretation of unseen spiritual truth.

Law is also feminine. It attempts to express the degree of unseen ethics of the masculine gender.

All works of art, whether written, spoken, painted, sculpted, acted, or sung are material manifestations of the unseen. A study of art is a study of the unseen message the artist attempts to express through the limitations of material media.

The path to ultimate truth travels through the realization that comes from an awareness of the unseen.

MAXIM 9

The "I" of the personality reflects the masculine gender and the "ME" reflects the feminine gender.

The masculine and feminine genders operate within as well as without. In humans these principles manifest as the "I" and "ME." The "I" functions as the assertive influence and the "ME" functions as the receptive entity.

The "ME" deals with the feelings, emotions, ideas,

thoughts, desires, and hopes of the individual. The "I" provides the motivational force to bring the feelings of the "ME" into reality. Western society has become increasingly "ME" oriented and depends more and more on an outside source to bring its desires and hopes into reality.

The "ME" can be changed through the conscious effort of the "I" or by being receptive to outside influences. Those with a well developed "I" attain their objectives whereas those with a weak "I" become victims of circumstance and happenstance.

The interaction between the assertive influence and receptive entity takes place within the body as it does in the universe, for the body is but a microcosm of the universe.

MAXIM 10

The masculine and feminine genders have characteristics that serve as attributes enabling them to fulfill their respective functions in the propagation and preservation of the species on its infinite journey of spiritual growth.

Masculine attributes are: abstract thinking, activeness, ambition, constancy, contemplation, courage, creativity, daring, discipline, force, independence, individuality, knowledge, leadership, originality, pioneering spirit, positiveness, progressiveness, spirituality, stability, and will-power.

Feminine attributes are: acceptance, accommodation, adaptability, caring, companionship, consideration, cooperation, diplomacy, friendliness, gentility, giving, harmony, industriousness, informative, materialism, nurturing, receptivity, responsiveness, rhythm, and visual thinking.

These attributes apply to all aspects of life. The feminine attributes of accommodation, adaptability, receptivity and responsiveness require an environment of constancy in which to function.

The Sun provides the environment and means for Mother Earth to bring forth life and nurture it; males provide constancy as part of the secure environment in which females function.

Each attribute of one gender requires an opposite attribute of the other gender in order to function properly.

The visual thinking of the female, which enables her to nurture, requires the conceptual thinking of the male to provide the safe and secure environment necessary for her proper functioning.

Chapter III

REPRODUCTION, SPIRITUAL GROWTH AND ETHICS

The 10 maxims regarding the principle of gender are an arbitrary designation that could be reduced to five maxims or expanded to 15 depending on how one presents in material terms knowledge of the unseen. Western thought requires that all learning be expressed in material terms, which unfortunately keeps it shielded from an awareness of universal truth. The illustration of these 10 maxims in action will hopefully show why Western thought has these limitations and what can be done to free it from its material confines.

An understanding of these ten maxims can be acquired without reading the illustrations and explanations put forth in this handbook; a reflection on these maxims coupled with an observance of the functioning of the world could suffice. However, the materialistic and responsive feminine nature of Western thought inhibits its ability to reflect and ponder on matters unseen; hence physical illustrations and examples will be called into play in order to cultivate the understanding that produces knowledge.

Before proceeding with illustrations we need to first address the fundamental purpose of the physical universe, which is to provide a vehicle for the habitation and sustenance of the soul in its present state of spiritual development. The soul has been given a material abode called the human body and a place in which to live called the universe.

The primary activity of the universe and all that it contains is the propagation and preservation of itself. If any aspect of life ceases, that activity will perish. The entire universe is in motion and that motion is directed at reproducing itself. Gender makes this activity possible.

An important component of human interaction is behavioral conduct. While most of life operates on instinct human beings have been endowed with a superior mind that enables them to function below, at, or above the instinctive level. This freedom of action can produce complete chaos or complete bliss, depending upon the values of the people and the societal structure in which they live.

Patriarchy is the natural societal structure and it contains the values that develop the ethics necessary for high moral conduct. Men who recognize that their purpose is to provide the environment and means for women to bring forth life and nurture it develop the ethics upon which moral conduct depends.

I ask that you remember the source of ethics. Earning money for its own sake has no ethics or morals; it is an aberration of natural behavior. Societies based on earning money rely on laws to maintain order and safety in an attempt to compensate for their lack of ethics and morals.

CHAPTER IV

MATERIAL FOCUS OF WESTERN THOUGHT

The previous chapter indicated that the primary human activity consists of propagating and preserving the species while on its journey of spiritual growth; a truism necessary for the understanding of gender and the patriarchal structure through which it operates. Western thought on the other hand, believes in the totality and finiteness of the material world and the right of the individual to live an unrestricted self-centered life in pursuit of pleasure and material accumulation. This difference in the basic understanding of human existence and purpose causes Western society to be out of harmony with the universal structure; it also reflects the feminine, materialistic focus of Western thought and inhibits its ability to see otherwise.

The knee-jerk reaction to the last statement is to ask the question "What is wrong with feminine?" There is nothing wrong with either the masculine or the feminine, but when they are out of balance a lot goes wrong. The Sun and the earth have a balanced relationship, as do the electron and the nucleus of the atom. In every universal relationship a gender balance must exist or the relationship will malfunction until balance is restored.

To attempt to have Western thought change its view requires formidable effort while the chances of success are extremely remote. Western thought has created a vocabulary, method of inquiry, and criteria of "proof" that reinforces its materialistic views and precludes any changes to it.

In the course of presenting my views to a college professor via Email I was informed by her that I didn't offer any empirical data to support my thesis. Empirical data? What does that have to do with ascertaining the truth? The reliance on empirical data represents Western thought's approach to "prove" things by the observation of material phenomena that can be reproduced and measured. Truth is not provable; it is understood. Also, I do not offer a thesis; I describe as best I can that portion of the truth that has been revealed to me.

Western thought continually searches for the truth but never finds it because it is looking in the wrong place. It looks outside into the material feminine world. The truth is found inside, in the unseen masculine world. "My kingdom is not of this world," "The kingdom of God is not in one place or another, the kingdom of God is within," "Man does not live by bread alone," "What does it suffer a man to gain the whole world and lose his own soul," are great mystical truisms not understood by Western thought. Consequently, it languishes in the ignorance of the material world.

The next chapter will describe the relationship of the Sun to the Earth, which will illustrate the interdependence of the feminine and masculine gender.

CHAPTER V

THE RELATIONSHIP OF THE SUN TO THE EARTH

The Sun provides the environment and means for the Earth to bring forth life and nurture it.

In order to enable Mother Earth to fulfill her function the Sun holds her in orbit, moves her through the universe, and provides life-sustaining energy. Mother Earth cannot produce without the unseen influence of the Sun. The Sun cannot foster the creation of life without the physical presence and activity of Mother Earth.

All gender relationships mirror the relationship of the Sun to the earth. To think and act otherwise courts disaster; a course Western society has chosen.

The idea of developing the feminine side of men and the masculine side of women to create balance is erroneous and out of sync with reality. Balance exists when the masculine gender does what it is supposed to do so that the feminine gender can do what it is supposed to do. To make the Sun more like the Earth and the Earth more like the Sun will result in the collapse of the solar system. Likewise, Western society is collapsing because of its genderless activities.

If people understood the difference of what each gender does, they would not try to equate behavior. For example, in the early days of American film there existed a popular ballroom dancer by the name of Fred Astaire. One of his more famous partners was Ginger Rogers, who upon her retirement from dancing said, "I did everything Fred Astaire did only backwards and in heels." This statement reflects pure feminine thinking. What she meant was that she thought she could do what ever she "saw" Fred Astaire do. However most of what men do is unseen and therefore beyond the understanding, let alone the capability, of what women can do.

Fred Astaire, decided where to bring Ginger Rogers, he figured out how to get her there, and then provided the energy to do it. His relationship to Ginger Rogers was the same as the relationship of the Sun to the Earth.

This gender relationship is so ingrained in our behavior that it is taken for granted and not even recognized as such.

When a man asks a woman if she would care to go to a rock concert on Thursday evening at 8:00 PM and she agrees, he then lets her know what time he will pick her up. He does the same three things that Fred Astaire did in his dancing; he decided where to bring his date, how to get her there, and provided the means to do it. This activity is expected, and if the man doesn't do it to the woman's satisfaction she doesn't go out with him again.

We will see this expectation in more detail as we address mating rituals in the next chapter.

CHAPTER VI

MATING RITUALS

To better understand the female and male requirements of their mates I will describe the mating ritual first of the doe—the female deer—and her prospective mate the buck; then that of the lion.

The doe, after having captured the attention of her suitor will turn tail and run away. If the buck follows she will increase her speed, then run over treacherous terrain, and perhaps jump across a wide ravine. Periodically she stops and circles the buck as if signaling approval but when he attempts to mount her she runs off again. Eventually she indicates her approval and then they mate.

I have told this story at lectures and I am continually amazed at how few young women have any idea of why the doe behaves as she does. They look at the ritual strictly in the romantic sense expecting that the buck will really be in love with the doe. For what purpose? Do they expect that the buck will get her flowers for her birthday and chocolates for Valentines Day?

The doe is testing the agility, swiftness, stamina, and perseverance of the buck. She knows that once she mates with him she will be in a vulnerable position and needs to be protected. She expects that he will have the ability, courage, and desire to protect her. She expects the buck to do what was described in previous chapters, to provide the environment and means that will enable her to bring forth life and nurture it. More than that, she expects that he will also be willing to die for her.

Like the doe, a woman should select a mate willing to die for her. However, the modern Western woman actually believes she doesn't need a man to protect her. She has the government, laws, police, surveillance cameras, security guards, and army to provide for her safety. One day soon, this government will fall and all protection will disappear; to whom will a woman then turn to for protection?

She will do what women have done from the beginning of time—mate with the biggest wielder of a club she can find, or someone able to have a few big club wielders working for him. This is the meaning of the words contained in Isaiah 4:1, "and seven women will approach one man and say to him we will make our own clothes and provide our own food, please give us your name." After the completion of the fall women will be lucky to find one real man for every seven women.

Now that we have finished our look at the mating habits of the doe, we will review the mating habits of the lion. The lion doesn't chase after females; he is the king of the jungle and females come to him. During mating season some lions mate with scores of females. After mating the females usually travel about on their own and then return to give birth and raise their offspring in the protective and cooperative environment of the pride. If the lion recognizes the scent of the lioness when she returns to the pride to deliver her offspring the other females will assist in their care. However, if the lion does not recognize her scent the cubs will be killed and the lioness driven off.

When I tell this story at lectures I also find little understanding of the lion's behavior. The lion has the cubs killed if he does not recognize the mother's scent for two reasons. The first has to do with the fundamental purpose of mating. The lion wants what comes out of the lioness to be a result of what he put into her, not of what some other male put into her. The second reason derives from the first; the pride cannot be kept disease free if the females mate indiscriminately. Every female in the pride mates only with the lion that heads the pride.

The lioness and her offspring are the focus of lion society, as are does and their offspring of deer society. The female mated with the strongest masculine male available so that she would have the means and the environment in which to bring forth life and nurture it. The male sets and enforces the standards of behavior for the societal grouping. This joint activity makes for a healthy and natural propagation and preservation of the species.

Unfortunately, Western thought operates in a manner diametrically opposed to this natural order. The suppression of natural male authority has removed from society all standards and their enforcement resulting in complete moral decline, mutation, weakening, and eventual dissolution of the species.

In the next chapter we will address the functioning of natural societies.

CHAPTER VII

VILLAGE LIFE

The village has served as the natural habitat for humankind for millennia. This chapter will address the gender relationship of a hypothetical Plains Indian village in America and a village in central Africa before the arrival of Europeans to both lands.

Activity in the Indian village centered about the care and well-being of its inhabitants such as the feeding and clothing of all, the training and development of children, and the tending to the needs of the aged and infirm. The women cooked, cleaned, sewed, mended, washed, nursed, and nurtured.

If the women determined that buffalo meat was running low and that junior needed a new loincloth, they advised the men of their needs. The men would then arrange a hunting party in order to obtain the meat and skins necessary for the nurturing activities of the women.

Pre-Columbian America had no horses or rifles, and hunting for Buffalo on foot armed with a bow and arrow had little likelihood of success. The method used to hunt buffalo consisted of getting a small number of the herd to stampede off a cliff. This activity required the coordinated skill and discipline of the group. If the operation went well there was plenty of food and material for the tribe. If the hunt didn't go well a few braves might be lost, and a few squaws would be looking for new mates.

Activity in the African village also centered about the care and well-being of its inhabitants. Women cooked, cleaned, sewed, mended, washed, and nurtured. They also did some gathering, and on occasion farming. In a tropical climate food is reasonably abundant and clothing demands are minimal; however, safety from the attack of an occasional marauding man-eating lion or tiger posed a significant challenge. If such a cat was spotted in the area or had been known to molest a neighboring village, the women would refuse to gather or farm and called upon the men to do something about their perceived lack of security.

Hunting a lion armed with only a spear is a relatively one-sided affair, with the lion having the advantage. Adding to the lion's edge is his preference to travel at night when his vision exceeds that of man's. To rid the area of the lion men established hunting parties made up of the most capable and courageous warriors. If they were successful, the safety of the tribe was restored as the marauding lion was eliminated. If the hunt didn't go well, a few warriors might have been lost and a few women would be looking for new husbands.

In both of the cultures described, the fulltime activity of the men consisted of providing the environment and the means for women to bring forth life and nurture it even if it meant giving up their lives. The environment in which women functioned was completely provided for by men.

The social mores and customs of men dealt with preparing themselves to better provide the environment and means for women to do their nurturing. Men naturally evaluate each other to determine where they could best serve the organization. Sports provided an opportunity to assess the physical prowess and courage of the men. It also gave the women an opportunity to evaluate prospective mates.

In the next chapter we will address the gender related productive environment of the modern American woman.

CHAPTER VIII

THE MODERN WESTERN PRODUCTIVE ENVIRONMENT

The city, or a combination of city and suburbia, provides the most common habitat for modern society. Large businesses or government centers of operation along with their satellite locations have become the primary method of providing goods and services. However, the male-female relationship in productivity remains unchanged from village times, though unrealized and obscured, especially as it regards direction and movement.

If any grouping of women were asked to raise the left hand and point to a compass direction such as North or West there would be few that could do it. If the direction uptown or downtown were added to the request the likelihood of error would be even greater.

Women have difficulty with directions because they cannot be seen. Not only are north, south, east, west, left, right, uptown and down town, unseen, but they do not even exist. They are arbitrary conceptual designations of men, as are all indicators of direction and location such as latitude and longitude, the Equator, the Arctic Circle and The Tropic of Capricorn.

Without the conceptual designations that men have devised, movement from one location to another would be impossible. Directions would be limited to going down to the river, up the hill, over to the lake, past the fir trees, or next to the pasture, which is the way women give directions.

Furthermore, the rate of movement is predicated upon unseen and non-existent designations such as second, minute, hour, day, week, and year. Also measures of distance such as inch, foot, yard, meter, mile, and kilometer are all non-existent arbitrary designation made by men. The movement of society throughout the world is predicated upon the unseen, non-existent designations established by men. The entire material world functions as a result of the unseen conceptual thinking of men.

The products that we take for granted such as the airplane, automobile, railroad, television, radio, telephone, radar, sonar, computer, and copiers all came into being as a result of the conceptual thinking of men. Modern Western women are as dependent upon men for what they have as women have ever been.

Now let's take a look at the workplace where the modern American woman feels that she can do whatever a man can do. Let's go to the assembly line where through union contracts and government legislation gender designations have been eliminated. Women will say that they are doing the same work that men do; they aren't—men are doing the same work that women do and that is not only the big travesty of Western thought, it has created a huge tragedy as well.

Men on the production line don't even do the equivalent of what the Indian squaw did when she tanned hides and made clothes. At least she had the responsibility and control for the whole manufacture of making a dress. The African wife had the responsibility of growing the food, preparing it for cooking, cooking and serving it. The man on the assembly line just has a task. He doesn't produce anything. He can't see the fruits of his labor. He does donkeywork. He's a jackass on the treadmill of production. The women wanting equal opportunity in the work place can now join the jackasses as she-asses. The American labor market consists of jackasses and she-asses on the treadmill of production; a condition brought about by the ignorance of Western thought.

The seeds were sown for the demise of Western civilization when the ancient Greeks turned over to the city-states the care and well being of society so that they could pursue materialistic interests and activities. World society is now in the final stages of deterioration brought about by the ignorance and suppression of gender differences.

All societies deteriorate when they lose vitality. The source of vitality is the virility of men. However, the loss of virility also includes the loss of conceptual thinking upon which the material world depends. Western society is imploding as it neuters the source of its vitality and inventiveness.

CHAPTER IX

THE PROTECTION OF WOMEN

Women are designed to mate, become pregnant, and nurture the race.

The attributes of women that enable them to nurture the race such as accommodation, adaptability, and responsiveness make them vulnerable to indiscriminate mating.

To protect women and keep them chaste until mating time and to further protect them from salacious prey and debaucherous influences societies throughout the world since the beginning of time developed rules of conduct between the sexes.

Many religions also interwove the propagation and preservation of the species into their rites and rituals. Taboos against adultery, covetousness, incest, and infidelity abound in many religions as well as in most tribal societies.

The only natural reason for members of the opposite sex to come in contact with each other is for the purpose of selecting a mate and then propagating and establishing family.

In addition to providing for the better propagation and preservation of the species sexual mores, traditions, and structures provide an environment of security for women. They are safe growing up in their father's houses, living in their marital environment, and walking the streets in a gender respectful society.

Another benefit of these gender-oriented structures is that it levels up the character of women. They become more courteous, modest, discreet, prudent, caring, accommodating, respectful, virtuous, and nurturing to name but a few of an infinite number of attributes that women develop. They pass these traits on to their children and level up the race. Happy is the child that can call such a woman mother.

Women are the receptive entity of humankind. They will be what outside influences expect of them. That's why they are responsive to advertising, fashion, and governmental dictates. The natural influence on women comes from men. There has never been a society with moral men and immoral women or immoral men and moral women. Women will be what men expect them to be. Western man has no expectations for women and they are degenerating accordingly. The degeneration of women leads to the degeneration of children, which leads to the degeneration of the race.

The ignorance of Western thought concerning the purpose of human existence and the gender related activity necessary to maintain it has caused the downfall of society as we will see in the next chapter.

CHAPTER X

THE DISEMPOWERMENT OF WOMEN

The term "empowerment of women" has permeated the media, politics and education of this nation. It is a term that derives from the complete ignorance of gender abilities and relationships. The words power and empower are relativities. A power hitter in baseball gets more extra bases per hit than the average player. A powerful speaker moves people to accept a message more easily than the other speakers. A powerful car has more horsepower than most cars. To empower someone is to give them power relative to others.

To empower women relative to men in manly attributes is an oxymoron. A woman can never be empowered vis-à-vis a man in manly things. Men on the other hand, can be disempowered by castrating them, neutering them, and emasculating them, which has been done by law and their own ignorance. The result has not seen any increase in power of women—they are not hitting a longer ball nor do they have more horsepower—what has happened is that men are no longer permitted to exercise their power and the result has been a complete disempowerment of women.

The way a woman can be empowered vis-à-vis a man is in non-manly areas. When men's power was unrestricted they built families. They could not do this on their own because only women had the reproductive ability. Therefore, women were sought after and they became empowered because they could do something the men couldn't; they could bring life into the world and nurture it. This is called a "win-win" situation for women; no man can compete with being a good wife and mother. Being a good wife and mother was empowerment not at the expense of men but as a result of the power of men.

This female empowerment leveled up the entire race, because men set standards and women did their best to show that they lived according to those standards. Men married "the mother's of their children" and were very careful as to whom they would select for that responsibility. Women were treated with dignity. A myriad of courtesies were extended to women such as opening the door for them, allowing them to go first, and having them walk on the sheltered side of the sidewalk; they were protected by their fathers, brothers, and husbands. They were not treated as similars because real men knew that women were receptive and responsive and therefore vulnerable and required protection.

In the current era of the emasculated male women have become completely disempowered. They have been reduced to work units. They have become automaton workers and programmed consumers. All the natural protection given to them by men has been removed and their moral conduct has degenerated. Their natural attributes of compassion and consideration have been suppressed and been replaced by a cultivation of aggressive behavior. They have been recruited into police departments to learn how to shoot people and they have been recruited in the military to learn how to kill people. The nurturers of the race are being taught to become the destroyers of the race. They are being motivated to become single mothers, dump their children while still in diapers into state run child care centers while they get on the treadmill of production. They see their children end up in street gangs, on drugs, in prison, and shot dead in the streets.

The result of this aberration of human behavior is that women are breaking down. The number one debilitating illness of Western women is depression. They have become so "liberated" that they are going out of their minds, and at an ever-increasing rate. Having the government act as their husband is not working in their best interests. Only men can empower women. REAL MEN who know their responsibility to the family, tribe, and human race empower women.

In order for men to fulfill their responsibility necessary for the empowerment of women, they must understand the principle of gender and the patriarchal structure necessary to maintain the family. They must also have the courage to regain the authority necessary to fulfill their responsibility.

CHAPTER XI

THE DEVELOPMENT OF OUR CHILDREN

The purpose of humankind is to propagate and preserve the species while on its infinite path of spiritual growth. To accomplish this objective requires the natural relationship of gender called patriarchy. While the mating urge itself is instinctive, the characteristics necessary to bring about the propagation and preservation of the species require development. This development begins with the birth of the child.

The very vibrations it receives from its environment as it leaves the womb has a developmental impact on the child. These vibrations affect its response to the nurturing care that it receives from its mother. Mother's milk flows freely when she feels safe, secure, and contented. It does not flow well when she feels tense, nervous, and apprehensive. Providing a safe and secure environment is the responsibility of the man. Filling this environment with nurturing love is the responsibility of the woman. Being a good wife is part of being a good mother and being a good husband is part of being a good father. The contentment and well-being of the child is a function of the level of patriarchy that exists in its environment.

When I enter the home of a person I can feel its aura as I cross the threshold. Just imagine how much more sensitive a child who not having fully developed physical senses and intellect is to this aura. The aura of the home is the life that the child feels 24 hours a day. Whether the aura is contained in a tepee, hut, house, or mansion is irrelevant; the early child does not have cognizance of structure, but only of its feelings. All who reside in and visit the child's environment—whether parents, grandparents, aunts, uncles, siblings, and cousins—contribute to the aura that the child responds to. That grouping of people is known as the extended family or tribe and is the natural societal grouping.

Compare the loving and caring environment provided by the extended family to the institutionalized care of people common in Western society. What kind of aura is the child enveloped in when there is extensive friction between the parents, when only one parent lives in the house, when the child is put into after school programs, when the child is in foster care, when the child is in an orphanage, or even when the child is cared for by a nanny?

The child requires the constant flow of loving care for its proper development. Loving care is made possible in a patriarchal environment in which all parties work for the common good of the family, tribe, and race.

As the child grows it observes behavior patterns and value structures behind those patterns. The child also observes the gender relationship between his parents. As the child matures the responsibilities of its gender are cultivated. Rites of passage were an integral part of most societal rituals in the non-Western world. Usually when children reached the age of puberty they were to a large degree removed from the influence of the opposite sex. This was especially true of boys who were then trained to fulfill their manly responsibilities to the tribe. When a girl in some of the tribes of the plains Indians reached puberty she was put in a separate hut to live in for a month with frequent visits from the shaman to prepare her for the most important function of her life—bringing life into this world and nurturing it.

Again, compare this way of life with the gross materialistic value structure of the Western world that motivates children to go to school, get an education, get a job, make money and be somebody. No mention of family, propagating the species, the tribe, or the race. Just make money and focus on the ME. It should be no wonder that one million of our school children are fed Ritalin, that America has the most violent boys in the world, that 40% of our children are born to unwed mothers, and that mental illness is the number one health issue of our nation.

Children of the Western world are being denied the natural structure necessary for their proper development—the family. There is no substitute for family. Men make families. No men—No families.

CHAPTER XII

THE LACK OF NURTURING IN WESTERN SOCIETY

I have repeatedly stated throughout this book that the purpose of the masculine gender is to provide the environment and means for the feminine gender to bring forth life and nurture it. Western society rarely uses the term nurture. Instead, it uses the term care and refers to those in need of it as children, aged, indigent, homeless, poor, under-privileged, hungry, uneducated, sick, infirm, orphaned, handicapped, and jobless.

The "care" expressed for others in Western society focuses on the need to sustain them physically. Requests for care for the needy can be heard in the rhetoric of those in the pulpit, in politics, and in various humanitarian and community organizations. The solution recommended by all of them requires contact with the appropriate institution that has the facilities and structure to provide for the specific needs of those requiring help.

The institution-oriented nature of Western thought has made the institutionalized care of humans a cultural export. Day care centers, orphanages, old age homes, mental asylums, homes for unwed mothers, and homeless centers are all of Western origin.

The institutions that have the most complete range of services, which include the provision of food, shelter, recreational facilities, medical care, vocational opportunities, and educational facilities are prisons; however, prisoners do not seem to be particularly happy. On the other hand, people on the outside of the prisons do not seem to be happy either. Depression is running rampant for people inside the cages and outside of them. Could the reason be that the institutionalized care of people is inhuman and unnatural? Could it be that institutionalization—which is a prison mentality—does not make for happy people?

The gross materialism of Western thought has reduced human needs to material considerations even though the great Master has stated clearly that man does not live by bread alone. Nurturing can only begin when society understands that truism.

Prisons, army posts, naval vessels, universities, and institutions provide for material needs and desires, sometimes better than the home does, but still we all yearn to be in the home. The home contains special attention transmitted by a loving wife and dedicated mother to all within her jurisdiction. She transmutes spiritual love into physical expression. She infuses all her actions and activities with unconditional love. When she cooks she adds an ingredient to all her food not found in stores; it's called love. When she prepares a meal she envisions the effect it will have upon those whom she loves. When she does the laundry she adds love to every fold. When she makes the beds she tucks the sheets with love. This special attention is called nurturing.

Women who dedicate their lives to nurturing tend to be very happy, and free of severe emotional issues. They exude love and in turn they are filled with divine love. They create a loving aura in their homes that draws members of the family to it. The warm and caring environment of the home affects all who reside there and helps them to live a more relaxed, productive, and caring life.

The nurturing woman needs a safe and secure environment in which to function; mother's milk can only flow freely when she is safe and secure. Family provides a secure environment.

Unfortunately mother no longer has a safe and secure environment in which to function. Government has eliminated the authority of the natural provider of a safe environment—man. Women now are "independent" and on their own. They can work, vote, own property and provide for themselves.

The children whom they once nurtured have been turned over to the state for "care," not nurturing. Her children live in an institutionalized environment for most of their childhood and are conditioned for further institutionalization upon adulthood.

These conditions make for unhappy women with the result that their primary health condition is debilitating depression. The children raised in institutionalized care aren't doing well either. America now has the most violent boys in the world, which make them prime candidates for penal institutions, where close to three million men now live. Three million girls now suffer from depression and close to one million grade school boys take Ritalin on a prescribed basis. The World Health Organization announced that the number one health issue of England, the United States and Canada is mental illness. The institutionalized non-nurtured world of Western society isn't functioning very well.

There is only one species of human life designed for nurturing; it's called woman.

The only natural institution for a woman to live in is called family. Men make families. Men with authority commensurate with their responsibilities make families. Male authority derives from the natural patriarchal structure of the universe. No patriarchy—no real men. No real men—no families. No families—no nurturing.

CHAPTER XIII

ATTACKING THE SACRED SHIBBOLETH

This handbook has shown that the propagation and preservation of the species while on its infinite path of spiritual growth requires the cooperation of two dissimilar entities: the masculine gender and the feminine gender.

By what reasoning then does Western society consider men and women to be similar?

By what reason does it employ the equal opportunity provisions of the voting rights amendment to apply to relations between men and women? By what reasoning are men's clubs forced to admit women and then give them the opportunity to vote on matters that on their own they would never address? A gross ignorance of gender and the purpose of human existence accounts for these actions. A sinister force also makes use of this ignorance and will be addressed later; however, even without the presence of this sinister force, the damage caused to the race by the almost inconceivable lack of understanding concerning gender and the purpose of human existence exhibited by Western thought is devastating and traumatizing.

Let's first take a look at what men do beginning with their early boyhood. Boys form clubs, teams, gangs, and organizations. They evaluate each other's talents and determine where those talents would best serve the organization. Boys and then men form inclusive organizations. They include as many members as they can in anticipation of the need arising for a particular ability. They practice discipline and competitiveness in order to maintain maximum effectiveness. As adulthood is reached these organizations evolve into hunt parties, fishing expeditions, and protective patrols. Their ultimate purpose is to assist in providing the environment and means for women to bring forth life and nurture it. Men are uniquely qualified to build these organizations through mental, physical, and spiritual attributes.

Now let's take a look at what women do beginning with their early girlhood. Girls form cliques. They begin forming them in pre-kindergarten and continue through grade school, high school, college, and any organizations they belong to for the rest of their lives. The formal name for clique is aristocracy; it is an exclusive organization. There is a function for the aristocratic feminine nature, but only if it operates within the confines of patriarchy. Women do not form clubs, teams, gangs, and organizations. They do not build—they exclude.

Every so-called woman's organization in one way or another depends upon the masculine influence. The PTA depends upon the school system that men created; nuns and convents depend upon the priesthood; women's business organizations depend upon the businesses that men created, and women's professional organizations depend upon the professions that men created. The preponderance of non-profit organizations that women belong to depend upon government law and the endowment policies of charities for their existence. Women do not have an equivalent of The Benevolent Purple Order of the Elks. The aristocratic nature of women does not lend itself to building organizations among themselves. The few that develop do not endure.

In February 2007 I posted a blog on mensnewsdaily.com entitled *Why Women Operated Businesses Fail,* which received a response from a blogger indicating that a study made in England concluded that not only do women operated businesses fail, but that all completely female organizations fail. Of course they do, organizational structure, comraderie, and purpose are not part of the feminine make up. An all female organization cannot sustain itself anymore than a solar system can sustain itself without a Sun.

Why then do women vote?

Because government has decreed that it is their right. Why? By what right do they vote? All organizations have qualifications for membership and the right to vote within them. A member of the AMA completed residency and received a license. A member of the ABA passed the Bar exam. Members of the Airline Pilots Association fly planes. What qualifies women, who do not have any organizations on their own, to vote in organizations that men have created? Lacking qualifications they were given a "right " to vote. What is a "right?" Society requires responsibilities, obligations, and considerations in order to function properly. There are no natural "rights." Rights are a fabrication of Western man to make up for his lack of understanding of patriarchy and to provide a vehicle for government intrusion into private affairs. We will see in the next chapter the part played by "rights" advocates in enslaving society; this chapter will maintain its focus on the effects of voting women.

Once women received the right to vote the masculine influence began to be watered down in the largest male organization in America, the voting public. As we have seen previously, women cannot become empowered vis-à-vis a man; however, men can be disempowered, which is what happened when women started to vote in the political arena. The effect of the women's vote was modest for the first 40 years due to the relative lack of feminine organization and the stabilizing influence upon society of the Great Depression and then World War II. However, the Great Depression and World War II planted the seeds for the further deterioration of male influence in society and especially in the family. Dad wasn't home very much during the depression years; he usually worked six days a week and was often gone from the house 12 to 14 hours a day. This left dad time to do little more than eat and sleep at home. As the depression rolled into the war, increasing numbers of men left the home to enter the military with 16 million American men having served in that conflict by the time it ended. While society at that time was essentially masculine gender oriented, it was the Queen who ruled the castle.

For more than a generation America and most of the world saw the home relatively void of the masculine influence. This is the equivalent of functioning without the yang, which Chinese philosophy explains will destroy the yin as well. That arrangement has produced the Tyrant Mother, a normal woman put in the un-natural position of raising a family without a competent masculine presence. Their children normally under-perform and also have difficulty in developing close relationships with the opposite sex. See essays titled The Tyrant Mother and Children of the Tyrant Mother in the essay section at the end of this book.

That generation raised by a tyrant mother spawned the counter culture of the 60's whose members were strong advocates of women's "rights."

As increasing numbers of laws were passed empowering women (disempowering men) in various aspects of societal activity the vitality of the various organizations that men had created began to diminish. Since women cannot act as men the level of conduct of the various organizations began to conform to female thought and its vitality decreased. As the vitality of the various men's organizations continued to deteriorate they became increasingly ineffective. To compensate for the lack of effectiveness of what had once been men's organizations, government stepped in. The 1960's ushered in Great Society legislation, the effect of which was to set the stage for the extending of the tentacles of government into every facet of societal activity. This legislation set the mechanics in motion for the final destruction of the family and the resultant elimination of ethics necessary for moral conduct.

Other chapters in this book and the supporting essays detail the elimination of ethics and the lack of morality inherent in every aspect of society; however, I feel the effect on the destruction of the family as a result women's voting (the disempowerment of men) is germane to this chapter.

In the 1965 New York Senator Patrick Moynihan wrote a report in which he elaborated on the problems faced by African-Americans in attaining equality of success. He pointed out the high rate of unwed motherhood, which at that time stood at 25% and indicated that the breakdown of the family inhibited their success as a group. Senator Moynihan called for a massive governmental effort to remedy the situation; however, many in the civil rights movement and in the African-American community derided his report. The government ignored his recommendations and emphasis was placed on quality-integrated education with equal rights in jobs, job-training programs and housing—all completely materialistic activities. The "equal rights" emphasis included women as well.

After 400 years of slavery and segregation the unwed motherhood rate of African-Americans reached 25%, a remarkably low figure and a testament to the inherent extended family culture in their society. After only 40 years of Great Society legislation the rate of unwed motherhood reached 70% and in some pockets such as Harlem it reached 90%. Harlem serves as an example of the de facto matriarchy resulting from the emasculation of the male. Nothing happens in Harlem without government intervention either overtly or covertly; a condition that cannot be considered progress for the African-American.

At the time of the Moynihan report the unwed motherhood rate in America was 3%; about double what it had been in the 1940's. After 40 years of Great Society legislation the rate reached 30% and as of this writing it has reached 40%. England, France, and Germany witnessed the same relative increases in unwed motherhood. In his last television interview Senator Moynihan was asked if he knew why the rate of unwed motherhood had increased dramatically in Western nations. His response was "No."

The brilliant Senator Moynihan who could see the absence of the male presence as a causative factor in the lack of equal performance among African-Americans could not understand what was happening to the Western world. He confused masculine presence with masculine authority. He too believed in the similarity of women. He too believed they should be given the right to vote and become a part of all organizations that men had created. For all his brilliance he suffered from myopic, materialistic, feminine Western thought.

The emasculation of the Western male brought about the destruction for the fundamental unit of humankind—the family. All ethics and moral behavior disappeared with the destruction of the family. The mystics teach as above so below and as below so above. The family is the smallest societal unit and its destruction has been devastating. The largest societal unit is the nation, and it too has become impotent.

To help illustrate the impotence of America I will include here an excerpt from my home page on the internet entitled The Land of the Smoldering Vagina:

Within five weeks of the World Trade Center incident, five different men and women said to me, "The phallic symbol of America has been cut off." I was surprised to learn of this undercurrent of feeling among the people, especially amidst the genuine concern for the victims and their friends and families, but upon reflection, I realized the depth of their statements.

The phallic symbol of America had been cut off, and at its base was a large smoldering vagina, the true symbol of the American culture for it is the western culture that represents the feminine materialistic principle, and it is at its extreme in America.

In 2006 and 2007 the New York City newspapers posted headlines "Still No Erection at Ground Zero," "When Will an Erection Begin?" and similar headlines. Ground zero symbolizes America. Vaginas do not get erections. America is the land of Viagra. As one women said to me "We can't get it up anymore." No we can't. Castrated, neutered, and emasculated men can't get it up anymore than women can. A disempowered man results in the demise of the family and the disempowerment of all men results in the demise of a nation. In an attempt to avoid the impending chaos the government has stepped into every aspect of private and public life.

There are only two options and one choice regarding the rule of society: Either men rule or the government rules. Western thought has opted for the latter. This decision, as we shall see in the next chapter, provided the environment for sinister forces to gain control of humankind.

CHAPTER XIV

THE SINISTER INFLUENCE IN WESTERN SOCIETY

Internet activity indicates that a growing number of people believe that sinister forces are actively engaged in the destruction of Western society. That viewpoint has some merit and it behooves us to determine what these influences are and where they came from.

Some believe that a socialist influence promoting feminist thought seeks to impose governmental control in all facets of Western society. Others hold that socialists and feminists are but tools of a conspiracy of international bankers who want to impose a one-world government.

The evolvement of an all-intrusive government based on financial interests is the natural outgrowth of a genderless society. What other kind of government could possibly evolve?

When people focus on material gain and self-aggrandizement they create governments devoid of ethics. These governments attempt to compensate for their lack of ethics by enacting a multitude of laws, which can never function as an effective substitute for ethics. Despite the laws of Draco, Solon, Justinian, et al, when male authority was eliminated in Greece and Rome extreme debauchery and immorality flourished until each regime collapsed. These same conditions have developed throughout the modern Western world and in other areas of the world where the Western culture has taken root. The environment created by Western thought spawns sinister influences.

Environments draw unto themselves those who feel comfortable in them, and also bring out in each individual responses to that environment. Positive environments produce increased spiritual behavior, and negative environments produce increased materialistic and self-centered behavior.

People do not live on a point of the spectrum of spirituality; they live within a span subject to environmental influences.

In a spiritually oriented environment people will tend to seek out mystical texts, they will find mentors and gurus, come across enlightened masters, and personal guardians. They will become aware of the saints that came before them, and all the company of heaven. They will realize their obligation, responsibility, and consideration to and for others.

In a materially oriented environment people will seek out texts on how to make money, they will find economic mentors and gurus, come across those whose lives are devoted to the accumulation of material things, and will draw unto them those who have gross carnal appetites. It is from this group that the sinister forces that operate in our society came; a negative environment spawned their existence.

Negative environments always foster the development of negative conditions and influences. A sitting pool of water will foster the development of a variety of microscopic plant and animal life. Unclean conditions in the home will foster the development of all sorts of vermin, and toxic conditions in the body will foster a host of diseases. These negative developments can be eliminated by changing the conditions that spawned them. Filtered and moving water will eliminate algae and other growth, cleaning the home environment will eliminate vermin, and proper food, exercise, and thinking will restore health to the body.

Eliminating sinister influences by force without changing the thinking that spawned them will recreate them. An illustration is the futile attacks on feminism. Conservatives and various men's rights groups are anti- feminist and believe if feminism were eliminated society would return to a more natural way of life. This is erroneous thinking. Feminism is not an entity unto itself; it is a symptom. It is not even a sinister influence; it is used by sinister influences to advance their objectives. If feminism were eliminated this very night, in the morning new feminists would be spawned because the thinking that created them had not changed. Western thought created feminism; it did not come from under a rock in the woods or from a hole in a tree trunk. Feminism cannot exist in a patriarchal environment; it cannot exist where people devote themselves to the propagation and preservation of the species and focus on the family. It cannot exist in a spiritual environment. Once the thinking that created the environment that spawned feminism changes, feminism will disappear as rapidly as vermin in a freshly cleaned home. It will have nothing to feed on.

The environment created by feminine, materialistic Western thought has produced many sinister influences aimed at control of the material world. These influences control the finances of this world and each year more and more people become enslaved by the economic system that they established. International banking exhibits the most flagrant abuse of power today; they literally own the world. To go into even the basic structure of how international banking operates would require not only more chapters, but separate books. Suffice it to say that they possess economic power heretofore unseen.

These sinister influences use feminism as one of many tools to help advance their objectives. Feminist lobbyists reach every senator, and representative in Washington. They reach the executive branch. They reach representatives and senators in state governments. They reach the governor. They reach mayors and city councils. They reach religious institutions. They reach industrial companies. They reach educational institutions. The cost of this extensive lobbying ranges in the tens to hundreds of millions of dollars annually. Where does this money come from? Did you ever see a feminist with a tin can asking for donations?

Sinister forces use women's liberation philosophy as a vehicle to suppress, diffuse, or eliminate completely male authority in order to create a breakdown of the societal structure and open the way for government intervention and then control. An example is the famous Virginia Slims advertising "You've come a long way baby," which implies that smoking expresses freedom from the domination of their husbands and fathers. Female smoking eventually increased the incidence of lung cancer in women, which started a 20-year campaign against the cigarette companies and the establishing of various laws limiting or prohibiting smoking. For the government to ban smoking at a cost of billions of dollars is called freedom but for husbands—at no charge—to prohibit the mothers of their children from smoking is called female suppression. The tragedy is that this viewpoint has been accepted.

A law designed for the ostensible purpose of helping women but in reality another vehicle for destroying male authority and societal structure known as Title IX calls for equal opportunity in education regardless of gender differences. The interpretation of this law considers males and females to be identical, a viewpoint contrary to even a rudimentary understanding of the purpose and function of gender. Boys and girls and men and women were put in the same classrooms and given the same instruction, which resulted in a drop in academic performance, unprecedented sexual activity, and an explosion in the rate of teenage pregnancies and unwed motherhood at all age levels.

Equal opportunity hiring has flooded the workplace with women in all labor classifications and management levels thus reducing job functions to non-judgmental tasks so that they can be performed by either gender. Also, as explained in Chapter IX of this book, the purpose of the sexes is to mate. By bringing both sexes in close proximity to each other whether in the work place, educational institutions, or various organizations, it broke down all societal safeguards regarding sex. It increased sexual activity, the incidence of adultery, divorce, unwed motherhood, and a disregard for the institution of marriage. What possible reason could there be for the forced sexual integrating of men's health clubs and gyms other that to increase licentiousness and decrease a sense of loyalty to ones' spouse? The primary support for all so called women's rights, empowerment, and opportunity organizations by sinister influences is for the express purpose of neutralizing male authority and breaking down the social structure so that governmental control could be imposed.

There is no such entity as a liberated woman, as an independent woman, or a self-sufficient woman. Everything in this world comes from men; there is no place else for anything to come from. The government has removed male authority and replaced it with government authority. It has set the standards for the care of children and if single mothers do not adhere to these standards the government takes the children and gives them to someone else who will raise them according to its dictates. Women are now dependent upon the government for child-care, healthcare, security, education, jobs, and old age care. Women are anything but independent.

Governments are dependent upon men for everything they have as well. They earn no money and create nothing. As their control of society increases its creativity decreases. The Soviet Union failed economically because it lost the ability to create NEW. North Korea will ultimately fail economically because it has lost the ability to create NEW. However, even without economic considerations all governments that do not support patriarchy will still fail because they do not understand the purpose and meaning of life and what is necessary to sustain it. They will ultimately implode. All Western nations are imploding; they have fallen out of harmony with nature and its dependency on gender; their attempts to fix what they have done continually meet with failure.

Gender is a natural metaphysical creation that enables the movement of the physical universe in which design denotes function. A penis and testis have a function; teats and a womb have a function; that function is to propagate and preserve the species. The mind and psyche of the individual are designed to be compatible with the physical design. A person is not a function of their gender, but gender is a function of the person.

The sinister influences in our society have fostered conditions that inhibit the natural expression of gender causing mutations to the race. When mutations occur in any species that affect its design that species begins its road to extinction. The polluted atmosphere (physical, mental and spiritual) of Western society is creating these mutations in gender and consequently is on the road to extinction.

In addition to the polluted physical environment the sinister forces have also polluted the bodies, minds, and psyches of the people. Having women participate in strenuous athletics is un-natural; it is not something that they have ever done on their own. Women are adaptable and responsive and they are engaging in these activities because they believe it is expected of them. These exercises are part of the mutation of the race. Women are developing broader shoulders, narrower hips and smaller breasts—three developments contrary to their natures.

Men's thinking mutates when they are educated in the same environment as women. They lose the ability to abstract, to conceptualize and deduce, causing them to be less effective in the propagation and preservation of the species. Prohibiting single sex men's organizations causes them to function in accordance with the feminine psyche with the result that most men are uncomfortable in them and drop out of membership.

Every opportunity for men to act as men has been blocked; the enemy has done its work well. It no longer has to publicly attempt to control society, the people are asking it to take control because they have lost the ability to manage themselves.

A review of the Ten Maxims of Gender and the illustrations indicated in the chapters of this book make it clear that there is only one assertive influence in the physical universe—the masculine principle. There is only one source of nurturing—the feminine principle. Living in harmony with the universal vibrations produces healthy, vibrant, and happy people. Living out of harmony leads to degeneracy and extinction.

CHAPTER XV

SPIRITUAL REFERENCES TO GENDER

Prophets, enlightened masters, and spiritual leaders since the beginning of time have traveled throughout the world teaching the people to be in harmony with the universe that God created. Scriptures of all religions describe activities beneficial to the physical and spiritual needs of humankind. Gender understanding, extended family, and the patriarchal structure necessary to maintain it are beneficial activities and are described in a multitude of spiritual texts; however, since this handbook has been designed for Western thought, Biblical illustrations will be used most frequently, even though these illustrations can be readily found in non-Biblical texts as well.

The purpose of gender is to provide functionality in this material world; it is not necessarily a determinant of spiritual growth although the patriarchal environment is conducive to spiritual development. Scriptures abound with illustrations of feminine spiritual attainment, but that attainment is reached by a path other than that taken by men and is manifested differently than it is in men.

Chapters VII and VIII of this book clearly indicate that all societal movement and all products come from the conceptual thinking of men. The Biblical allegory of Eve being born of Adam's rib illustrates this truth. Adam represents the masculine principle, the motivating force of the universe. All things and all movement come from the masculine principle, there is no place else for anything to come from. Also, have you ever wondered why the serpent spoke only to Eve? Adam represented the assertive masculine influence and Eve represented the receptive feminine entity. With Adam not around the serpent said, "Eve baby, I have a deal for you," and seduced her.

Many centuries later when the Magi saw a sign in the heavens that a master was coming, they got on their camels and rode to Judea where they inquired as to who was born about that time. When Herod heard about it he had all the males killed that had been born in a two-year period (about the time of the "star"). There was no doubt in Herod's mind that the Master would be a man, and he knew as the serpent did, that without the opposition of the male he had no insurrection to fear.

Revelations 12 contains the story of a woman pregnant with a male child and Satan in the form of a dragon waiting to devour it. The dragon knew that with the male child and its seed out of the way, there could be nothing to oppose him. Upon its birth the male child is snapped up and brought to heaven while its mother takes refuge in the wilderness as the fight between good an evil commences. The issue between good and evil or any other aspect of control entails the necessity of isolating the masculine influence and neutering it. There is no other assertive influence in the physical universe.

In 1st Corinthians Paul says Let the women keep silent in the churches...... If there is anything they desire to learn, let them ask their husbands at home. He says in 1st Timothy "Let the woman learn in silence with all subjection." The reason for these statements is that matters spiritual are unseen and conceptual, which is masculine thinking.

Paul also says The man is not of the woman but the woman of the man. This truism is also contained in the Koran, which states "He created you from a single being, then from that single being He created its mate." Both of these statements are in accord with recent scientific findings that only the male contains the Y chromosome, therefore the male cannot come from the female. The motivating force of the physical universe can only come from itself, there is no place else to come from.

Paul states further "I don't allow women to teach, nor do I ever put them into positions of authority over men - I believe their role is to be receptive." The feminine gender is the receptive entity of the universe.

From the first book of the Bible to the last book and many places in between we find examples of the principle of gender and a clear expression of the differences between the masculine and feminine manifestations. The Bible is a great compendium of illustrations of spiritual truths; is it reasonable to expect that over the 1500-year span of its authorship that the writers could all be wrong concerning gender differences? Isn't it more likely that gender is a fundamental principle of the universe and therefore references to it permeate this great spiritual text?

Various prophets have cautioned against the loss of male authority. The great Hindu epic The Mahabharata states "Woe unto the nation that has the tiara come before the crown." Mohammed is quoted in the Hadith as having said "A nation that gives its affairs to women will never be successful." Chapter VI of this book contains the words of Isaiah concerning women. The Chinese concept of the yin and yang indicates that without the yang the yin will cease to exist. The Hindus speak of the Shiva-Shakti energy (the masculine and feminine aspects of universal energy). The sources indicating the gender polarity of all existence seem endless. Only Western thought has difficulty grasping this concept; consequently Western society is imploding.

CHAPTER XVI

THOUGHTS FOR MY DAUGHTERS

You have been endowed with the unique ability to bring life into this world and nurture it. Each of you has the potential to be the Mother Earth of your environment. Mother Earth supplies our every material need and our emotional needs as well.

We have all experienced the calming and relaxing presence of Mother Earth as we sit under a tree in a field, on our lawns, in the park, or on the beach. We also feel that calming influence in the home of a loving and nurturing woman. The aura of her home contains the same loving and nurturing principle as the aura that we feel when on the beach or in the forest.

Mother Earth can only function if there is a father sun to provide structure, direction, and energy. The Mother Earths of the heavens have suns to rotate about; a relationship that exemplifies the natural patriarchal structure of the universe. The nurturing female love requires a structured and secure environment in which to function.

Society today suffers from a lack of the structured and secure environment necessary for the expression of nurturing love; consequently people feel lonely and forsaken. They attempt to escape from the angst that the absence of nurturing love creates through entertainment, sports, substance abuse, and sex, but these vehicles only provide temporary relief. Only a woman can provide the nurturing love that humanity requires, and she has been denied the environment needed for her to function properly.

Many women have been influenced to seek a "career" in lieu of marriage and family, or to supplement it. First let's look at what we call a career. Most all of the career jobs consist of activities created to adjust for an unnatural lifestyle.

Social work attempts to compensate for the lack of the family structure. Most therapies address the issues resulting from a lack of family structure. Our legal and criminal justice system's primary activities relate to people who come from broken homes. The pharmaceutical industry thrives on the unnatural lifestyle of society. The bulk of the medical profession deals with effects of unnatural lifestyles. Very little in the career field supplies the natural needs of humankind. The overwhelming activity of all Western oriented societies focuses on attempting to fix the effects of their unnatural lifestyle.

When we compare what we refer to as primitive societies or even less modern ones to Western society we find that they have lower rates of incarceration, suicide, mental illness, depression, crime, cancer, and illnesses. Once the Western culture collapses it will return to a more natural way of life and these career opportunities will be gone. They won't be needed. However, nurturing love will always be needed; humankind cannot long survive without it. Only you can provide it.

Financial independence serves as another motivation for women to seek a career. Earning your own money means that you also have the responsibility of providing for your economic well-being. This responsibility produces a major portion of the stress found in the modern Western woman. Why would you want that responsibility and the stress it produces?

In traditional marriages, even in the Western culture, before a father approved of his daughters engagement he wanted to be sure that the man who wanted to marry her would provide for her. Furthermore, there was the expectation in a marriage that a man would take care of matters financial and they would not be a concern of the wife. In rare cases when divorces did occur the woman was entitled to half the estate. Today, financially successful women pay alimony and child support to their ex-husbands.

The traditional wife with a responsible husband had an inner calm and an outward serenity that no longer exists in the modern liberated society. This state of mind enabled her to provide the much needed nurturing love of humankind.

Another topic worthy of your interest concerns the changing values regarding sexual promiscuity in Western society. It has been called the "sexual revolution" in part because of the overthrow of the natural constraints on sexual behavior imposed by various societies for millennia. Religious texts throughout the world and sexual mores of various ethnicities consider the chastity of women an integral part of the moral conduct of society. They also realized that the adaptive, receptive, and accommodating nature of women made them highly vulnerable to seduction. Social mores were instituted to protect women. The removal of these mores has resulted in the abandonment of women to the exposure to and influence of the lowest forms of moral and spiritual behavior resulting in devastating effects to all members and segments of society.

Men make standards—all standards. Real men recognize their obligation to the propagation and preservation of the species, which also includes its spiritual development. Real men establish and enforce standards that serve for the betterment of the social grouping in which they reside. The overthrow of these standards in modern day Western society is not a revolution, nor is it anything new. When men lost their influence in ancient Greece and Rome both societies degenerated into debaucherous environments that imploded. The modern Western world is experiencing the same conditions. There has not been a sexual revolution in the terms of enlightenment; we are witnessing a replay of what has gone on before and are experiencing the same negative effects.

If you want to experience the joy of fulfilling your natural feminine responsibilities it can only take place in a secure environment with the necessary means for you to function, which can only be provided by a real man. When you find that real man be obedient to him. Be obedient to your fathers as well. Both of these men love you. If they abuse you they will pay the price. You maintain your devotion, and remember that the spirit is always at work.

I share these thoughts with you in order to give you a basis to reflect upon other than the gross materialistic thinking of Western thought. The assertive influence of the universe is the masculine principle. Your government by limiting the authority of men broke down the family structure and made you dependent upon the government for your very existence. You have not been liberated—you have been enslaved. You have become harem girls of the state.

The only way out requires men to awaken, reassert their manhood, and recreate the environment necessary for your natural functioning. When you find the few willing to do it give them your support; it's your only positive option.

CHAPTER XVII

THOUGHTS FOR MY SONS

Since the masculine principle is the initiatory force of the universe it is fitting that the penultimate chapter of this book address those who possess that principle.

You my sons, and your forefathers have created all that exists on this earth, and are responsible for the condition of society. Depending on how you view society you can either take the credit or the blame.

Women are the receptive entity of the universe; they respond to what you initiate; their behavior is the behavior that you expect. It's time to stop blaming feminists or any political grouping for societal conditions and instead take a good look in the mirror.

In referring to the mirror I am taking a page from the therapist's book. They have two basic steps in changing ones behavior. First they get you to look in the mirror. People resist this first step because they usually don't like what they see. Once the first step is accomplished the second step consists of removing all the excuses for what one sees in the mirror. Once steps one and two have been accomplished the patient then can start making changes in behavior that will improve his life.

Addressing the problems created by Western thought requires taking a long look in the mirror. Secondly, a realization must be reached that Western thought created the conditions that produced the problems seen in the mirror. Once these two steps are accomplished then changes can be made to improve your lives and that of society.

What do we see in the mirror? We see a society degenerating in all respects (details have been given in previous chapters and in accompanying essays).

What is the cause? An imbalance of gender—specifically the lack of the influence of the masculine gender—has caused the degeneracy and subsequent implosion of societies where Western thought has taken root. Western thought, which contains an imbalance of gender, has not been in harmony with nature, the environment, and the spirit.

Many will say that men have always been in control in Western society and therefore question the assertion that it is feminine and materialistic. The assertive influence will always be the male, but the value structure that it operates from can be feminine. Western man turned over to the state the care and well-being of people so that he could pursue material interests. In so doing he created institutions that took over the responsibilities that he abdicated. Today these institutions control his life. He has grown so accustomed to these institutions that he goes to them for help in solving his problems, which is a natural feminine response.

Women instinctively know that the power is outside of them. That's why they scream when they're in danger. They make scenes with their boyfriends or husbands in order to intimidate them to provide what they cannot get on their own. When a group of women gets together and makes a scene it's called a demonstration but the intent is the same—to get a higher power to do for them what they cannot do for themselves. Now we find men demonstrating to their institutions. In doing so they are acknowledging a higher power than themselves—the very institutions that they created. Men now act as women.

You have brought about this condition.

You have legislated away your natural authority. You have enslaved yourselves. You have created the conditions for feminism to spawn. You have created institutions that involve themselves in and control every facet of your personal lives. You have created the mechanism for one world government. You have created the international financial system that now controls governments and indentures nations and people. You have reduced yourselves and your sisters to jackasses and she-asses on the treadmill of production while the institutions that you created control you.

You have done all of this.

You are destroying the environment necessary to sustain yourself. You have polluted the air, water, food, and thought of society. Western thought considers nature something to be conquered and tamed not something to be in tune with and a part of; therefore, it abuses the environment, uses it for self-gratification and refers to it as a resource. Western thought also considers people to be a resource and abuses them much the same as it abuses the environment.

Two and a half million of your brothers are rotting away in cages in the prison system. One million of your sons waste their youth as members of street gangs. You have allowed the family to be destroyed and left women and children to be cared for by the state. Only conquering armies could create and enforce such conditions and you have allowed it, permitted it, and even fostered it.

You have done this.

These are the conditions to be seen in the mirror. The next step—the determination of the cause of these conditions as already been stated—the lack of influence of the masculine gender and its inherent values.

A society needs balance in order to function properly. It requires the feminine influence to bring forth life and nurture it; the masculine principle provides the means and environment (including spiritual development) for this nurturing to take place. Western thought has feminine values. Western man is not aware of his spiritual nature. He functions in the material world. In order to make change he must change his thinking. He must be reborn. He needs to recognize the spiritual nature of the universe and everything in it. He needs to gain a sense of self that transcends the material.

A feeling of oneness with the universe and everything behind it needs to be developed. Man must identify with nature and with every form of life that he sees. He must eliminate his aristocratic thinking and love all people and see in them himself. He must see himself as an infinite spiritual being instead of as a finite material commodity.

When men adopt this thinking their values will change and their outer world along with it. Their activities will change from working for personal gain towards working for the benefit of others in both material and spiritual aspects. The need for institutions, authority, and laws will diminish. Instead there will be social mores, customs, and traditions that will serve to sustain people physically and nurture them spiritually.

The time has arrived my sons for you to become transformed and in the process you will transform the world. It will take courage to step away from the norm of society; however, always remember that you are the assertive influence of the universe; there is no power out there other than you. Once you come to this full understanding you will no longer need courage, for courage is the ability to do what you are afraid to do. Once you have the true inner knowledge you will realize that there is nothing to be afraid of.

The majority will not be with you because the majority never changes anything. Change is brought to the world by a small minority of zealots—fanatics—who have the courage to espouse and live their beliefs.

You have the power within to bring about all that you realize. Say to yourselves each day: I AM A MAN, IAM A MAN, I AM A REAL MAN. Then live your lives accordingly.

And remember, as you expect women to be obedient unto you, you must be obedient unto God. As this God does not abuse his power over you but showers you with knowledge, love, and understanding, you in turn must not abuse this power over women and children, but devote yourself to their well-being. That is a balance of gender and constitutes the patriarchal nature of the universe.

Be reborn my sons and shuck off the confining sheaths of Western thought. Go and change the world.

CHAPTER XVIII

SOME MUSINGS

I believe that what you have already read clearly stated, explained, and illustrated the principle of gender and the patriarchal life that it supports. In this chapter I am sharing with you various musings on the subject of gender and Western thought that might stimulate your psyches into further understanding and application of the subject.

Western thought cannot grasp the concept that the material or seen world is a function of the mystical or unseen world; it bases all of its knowledge (information) on what can be "proven" by material means. By making material proof the criteria in determining what it calls knowledge, it makes a god of matter.

In its determination of knowledge it presents a thesis based upon a premise supported by empirical data and/or test results. The nature of empirical data and test results is its materiality. Western thought has made a basic premise that all knowledge can be ascertained through material determination. It cannot conceive that the material world is a manifestation of the unseen world; nor can it conceive that all knowledge is unseen and requires understanding not proof.

The focus of Western thought on materiality precludes it from having any understanding of matters spiritual. Even Western philosophers and theologians from Plato to St. Thomas Aquinas, who might not have had much in the way of test results and empirical data available to them, nevertheless referred their deductive reasoning to a material basis.

Not having the capacity to understand spirituality Western thought has institutionalized its materialistic interpretation of the teachings of prophets and spiritual texts; it maintains its hold on the people through the establishment of authority, laws, and punishments.

Western thought does not know that humankind is on a spiritual journey; it believes that people are on a once through trip in a finite material world. Consequently, it has little concern with the propagation and preservation of the species, but great concern with the rights and privileges of the material self.

Western thought has little understanding of gender differences, as readings from Plato to Marx will verify. It believes that women and men are essentially the same except that men possess larger and stronger bodies on the average.

The agrarian culture saw a division of labor among the sexes in the Western world that approached natural patriarchal divisions, not as a result of understanding but as a result of survival. A pregnant or nursing woman was not about to chop down trees and build fences while her husband cooked dinner and watched the children.

The industrial revolution, production lines, and World War II combined to get women out of the home and into the labor force. Since the true purpose of gender was never completely understood by Western thought, once labor was simplified and made less strenuous people became work digits, and women qualified for this classification as well as men. The philosophy of "sex for pleasure" took root concurrently with the demise of the family.

Western thought has made a god of matter. Wherever Western thought prevails the dictates of its god Mammon are adhered to. Muslims worship in their mosques on Friday, Jews worship in their synagogues on Saturday, and Christians worship in their churches on Sunday, but on Monday morning they all face Washington and adhere to the dictates of the high priests of Mammon. All actions are subject to the scrutiny of Mammon's clergy. They have the power to breakup families, determine the curricula of educational institutions, approve or disapprove licenses necessary for ones employment, and supplant right and wrong with legal and illegal.

Boys are not groomed for the responsibilities of manhood nor are girls groomed for the responsibilities of womanhood; instead educational institutions train children to be genderless production workers. They are hooked to the treadmill of production early on and become enslaved due to the massive debt resulting from the cost of training that enables them to get on the treadmill.

The American Medical Association consists of true worshippers of Mammon. They treat symptoms, which they can see and measure, while they ignore causes. Treating symptoms requires medication—lots of it—so much so that pharmaceutical companies have grown into huge multinational organizations that use their finances to influence the media, manipulate public perception, directly affect medical protocol, and support a huge governmental lobbying effort. They have increased the cost of sick care to a level that government has had to take over its administration and funding.

Materialism has become ingrained in societal values to such a degree that when financial magnates, movie stars, sports heroes, lottery winners, and others of extreme wealth have domestic, emotional, and psychological problems society is amazed that with all that wealth they weren't happy. The god Mammon has its influence everywhere and it does not provide solace anywhere.

The unseen creates the seen. The unseen God created the physical universe and the unseen conceptual thinking of men created the material world upon it. The unseen attributes and authority of men have been neutered and the society that men had created has started to crumble. The seen cannot exist without the unseen. The physical universe could not exist without the unseen God (it is his creation) and the material world and its organization and structure cannot exist without the unseen conceptual thinking of men (they created it). The problems of society cannot be "fixed". A bad concept cannot be modified into a good one; only changed thinking can change society.

I trust that these musings will serve the purpose indicated at the outset and that they will motivate you to bring about change in your thinking and then in society.

CONCLUSION

This handbook states from the preface to the last chapter that Western society is imploding, which might appear to be at odds with a small but growing body of thinking that holds that we are entering an age of spiritual enlightenment. I too believe that the people of this world will experience a period of enlightenment, and trust that this handbook evidences a step towards it; however, before a person can become well, the illness within must be purged. This analogy holds concerning the health of nations as well. The thinking that caused the problems of a nation must be changed before its outer manifestations change. Western thought does not have the capability to consider that anything within itself requires purging; therefore, the society that it crated will collapse, and after that occurs new thinking will engulf society and the path to enlightenment will quicken.

The forecast of an implosion of society does not require any clairvoyant or prophetic abilities. Wouldn't reason conclude that a society that pollutes the very environment that it needs for survival will collapse? Wouldn't reason conclude that a society that has destroyed family and left its offspring to be raised by an immoral state collapse? Wouldn't reason dictate that a society bereft of spiritual understanding will drown in the excesses of gross materialistic pursuits and self-indulgent behavior? And lastly, could a society that does not understand the physical, mental, emotional, psychic, and spiritual differences in gender long survive?

The collapse of Western society does not constitute a doomsday prediction. Human life will continue on this planet; it always does. However, it will require the development of survival strategies. Learning to live minimally in all respects will be a major consideration, as will the necessity of learning to be in harmony with nature. Part of this harmony will occur when society by necessity develops the natural patriarchal structure of the universe as explained in this handbook. Men will recognize their natural authority and use it to create the environment and means for women to bring forth life and nurture it as humankind progresses on its spiritual journey.

ESSAYS

The following essays are based on the ten maxims of gender listed, explained, and exampled in this handbook. They cover a wide range of issues that affect our daily activities. As you read these essays you will begin to become aware of the gender influence of the universe and how the imbalance of gender in Western society created the negative issues affecting it.

Perhaps reading these essays that will cause you to rethink your value structure and its resultant lifestyle. A change in thinking produces a change in living. You have come this far; why not continue on the journey of positive change?

AMERICA HAS WORLD'S LARGEST PRISON POPULATION

America has more men in prison than any other nation; more than China and India combined and they have more than two billion residents among them. One man in six between the ages of 18 and 30 has a relationship with the law. In the African-American community one man in three in that age groups has a relationship with the law, with the trends worsening.

In 1976 the prison population numbered 250,000, a high figure by world standards. Eight years later in 1984 it had doubled to a half million, by 1992 it had doubled again to one million, and by 2000 it had doubled to two million. It now stands at 2.4 million and growing.

A society that locks up large numbers of its men in prison and has them spend what should be the formative or productive years of their life rotting away in cages, has an oppressive government. Yet, this issue never receives recognition let alone discussion from any political party or member of government at any level, nor even from the supposedly watch dog media. Those 2.4 million men in prison are someone's sons, brothers or fathers. That's a major issue.

There are all sorts of pretexts upon which to arrest males. Domestic violence, sexual harassment, and accusations of rape are but a few of them. When the new hate crimes legislation gets enacted to supplement these laws we will see a rate of incarceration that will approach the mass production of prisoners.

The government wants all male authority destroyed. It does this by a combined effort of neutering the male and incarcerating him. All laws, employment practices, educational instruction, and family matters are based on gender equality, which the law interprets as gender sameness—an illustration of the gross ignorance and materialistic thought of Western society.

However, in cases of domestic violence—where statistics indicate a true equality of violent acts—the police, who are now required to make arrests on all domestic violence calls, usually arrest the male. In cases of sexual harassment claims, the word of the female is usually taken. The foundation of law—innocent until proven guilty—is reversed in matters of gender issues and the male must prove his innocence. Gender sameness is not practiced in the legal system.

If the male does not comply with the un-natural practice of sameness he is thrust into prison, and as part of his parole or release he is required to attend support groups or counseling which in essence destroys his manhood.

The government has a purpose in promoting and enforcing gender sameness for it assists in destroying the authority of the male. For anything to function in this universe there can only be one assertive influence—one authority. Since by law gender sameness must prevail in the home that translates into a lack of authority. The government has stepped in and made of itself the authority in all matters concerning men, women, and children. In essence it has destroyed the family as the focal point of society leaving it leaderless and begging for the authority of the government.

Our ubiquitous and all pervasive government has brought us to the threshold of tyranny. It could only do this by emasculating and incarcerating our men—our fathers, brothers, and sons.

Will we stand by passively and become engulfed by tyranny, or will we rise to the challenge and join with like-minded people and re-establish among ourselves the natural patriarchal structure of the universe?

AN ALTERNATIVE TO WESTERN THOUGHT

I recently completed a list of 10 gender characteristics along with explanations of them and in planning on how to present them on the internet and to the public at large my thinking was influenced by two recent newsworthy events concerning family.

The first of these events occurred in Milwaukee, where social workers conduct strip and search procedures in private schools as well as public schools to determine if parents had spanked their children. They conduct these invasions of privacy on an ongoing and routine basis. The second event, which occurred a thousand miles away, witnessed Texas authorities removing 450 children from the parents of a polygamous group. These acts serve as but two of an infinite number of examples of the government's view whether on the national, state, or city level, that children belong to them and not to their parents.

These acts reflected the thinking of the Western psyche, and I planned to use those acts as an introduction to gender, but realized that I had already written an article titled The European Psyche, which I have since published under a Message from Elder George. However, an important consideration that I had left out had to do with the Platonic philosophy, which requires that the family unit have no place in the perfect state and that children should be turned over to the state in infancy. For the 2,500 years since the writing of the *Republic* the Europeans destroyed the family and cultural environment of every society they had contact with and have left in its place a monolithic government devoid of culture that has taken over the raising of children.

Organizations that believe in the importance of family blame communism, socialism, and feminism for the break-down of family, but those entities are merely manifestations of materialistic feminine European thought, which began in

Platonic Greece. Recognizing that what we see today is but an extension of that thinking, it is futile to make any changes within the system to strengthen family. A recent government statistic indicates that only12% of 18 year-old Americans live with both natural parents. For all practical purposes the family has become extinct and the state has taken over all functions of raising children.

The cause of the collapse and elimination of family does not lie with socialists and liberals, nor even with Platonic theory, but with the inherent value structure of Western man. He sees himself as a finite entity making a single trip through earthly existence. He therefore looks upon self-indulgence as his birthright. He sees the world as a huge ball of material that can be exploited for his pleasure. He lives in the here and now with no understanding of things unseen. He looks upon sex as a means of self-gratification rather than the vehicle through which the race is propagated; he has even developed chemicals to be ingested by women so that they do not become pregnant. The future of the race and world do not weigh heavy on his mind.

Western man does not see the opposite sex as being essentially different than himself, and since he considers the acquisition of material things, which he calls wealth, a prerequisite to happiness, he feels it only right and proper that women should have the same "right' to amass wealth. This "right" was expanded to vote and be involved in governance, an activity that the feminine psyche is ill equipped to handle. Since women comprise the majority of the voting population they have voted away male authority in the home, which resulted in the destruction of the family; in the educational system, which has eliminated standards and reduced our children to reciters of information; in the workforce, which has created a whore economy; and finally in government, which has resulted in a multiplicity of laws to make up for the lack of manly standards. Western man voted away his inherent ability to rule and reduced himself to an automaton production worker—to a jackass on the treadmill of production. The thought that he did this of his own free will is a testament of his massive ignorance regarding his purpose in life. When Jesus uttered the words, "Forgive them for they know not what they do," he was speaking to the ignorance of Western man that had permeated the area for more than three centuries.

A view that serves as the antithesis of Western thought looks upon the purpose of mankind as one of spiritual growth and its primary activity to be the propagation and preservation of the species. It requires that he be in tune with nature (God's creation) rather than attempt to control or alter it. The structure through which this activity is accomplished is called patriarchy, a concept not understood by Western man. As is his wont, Western man requires definitions of all that is not natural to his way of life. Wikipedia offers the definition of patriarchy as "The structuring of society on the basis of family units, where fathers have primary responsibility for the welfare of, hence authority over, their families." That's a satisfactory workable definition, but I offer a simpler and more universal one.

Patriarchy is the natural interplay of gender. That's it. No parliamentary decisions to be made or legislation enacted. No empirical data to be amassed; no theories to be proven. It occurs in the solar system and in the atom, and in everything in between.

AN INABILITY TO MAKE CHANGE

The seen and the unseen comprise all existence. Not that some things are seen and some unseen, but that all things seen are manifestations of the unseen. The visible abundance of mother earth represents the unseen energy of the sun and the unseen effect of gravity that holds her in a position for optimum functioning. The ebb and flow of the tides reflect the gravitational pull of the moon. Religious law is a manifestation of the interpretation of spiritual truth. The visible world is an effect of the unseen world. An unseen causative force accounts for the condition, appearance, and movement of things seen.

The ancients designated this unseen causative force the masculine principle; it motivates and affects the feminine material world. The masculine principle provides the environment and the means for the feminine principle to bring forth life and nurture it. All existence operates under this principle. The feminine principle cannot of itself bring life into this world, nor can it properly nurture without the unseen parameters of the masculine principle. Any society that minimizes the effect of the masculine influence deteriorates. The fall of Israel to the Babylonians, and the declines of Greece, Rome, and Spain exemplify the effect of the decline of the masculine influence.

I have discussed the effect on societal conduct by the weakening of the masculine influence in other writings, the purpose of reviewing the nature of the masculine and feminine principles here is to illustrate how all the issues of Western society remain unresolved because it works on what it sees, instead of on what is unseen. As an example, the attempt to correct poor academic performance focuses on reducing class size, providing modern teaching equipment, and paying the teachers more. None of these attempts have worked because the underlying unseen issue was not addressed; the breakdown of the family, specifically the absence of the father at home and the unseen values that he would normally impart to the children.

Nowhere is the inability to address causative effects more entrenched than in the medical and dental professions. The entire medical profession has designed a protocol to treat symptoms instead of causes, consequently the population has become increasingly dependent upon medication that treats symptoms but does nothing to address causes.

We are in the throes of a health crisis in America because we do not know the cause of illness. Heralded medical and scientific advances have to do with reconstructive surgery and dentistry; activity more in keeping with Western thought for it approximates constructing bridges, tunnels, and buildings–material things; however, many times these surgeries must be redone because the underlying cause was not addressed. Also, most surgeries are unnecessary. In any event, surgery occurs not because of good health but because of poor health.

One of the big expenses of modern medicine is the research necessary for the development of "cures." Effects are not curable, but causes can be altered or eliminated.

What medication cures anything? All medications treat symptoms. Western man gives these symptoms realty, but they are only effects; when the cause is altered or eliminated the symptom disappears.

In the recent publicity of the humanitarian activities undertaken in Africa by Bono and Mr. & Mrs. Bill Gates the media indicated that one person in Africa dies from Malaria every 27 seconds. I discussed this with a Kenyan woman who told me that when she was a girl malaria was cured within 48 hours by the use of herbal treatments. She came from a family of 12 children, and her mother raised another 12 and none of them died from malaria. She said the health problems developing in Africa arise from the breakdown of the immune system caused by eating Western foods. The first thing Westerners gave to her people was chocolate. Then they were told to eat white bread instead of the foods they normally ate. As their diet became Westernized their immune system could no longer stay in tune with the environment in which they lived. Then Western man came in with his medications to "fix" things. Her viewpoint was that Western man goes around the world destroying the natural order of things and then attempts to fix the symptoms of this destruction. I thought that was well put.

The air, water, food, and thinking in America have been polluted and we attempt to "fix" it with drugs, surgery, therapy, and incarceration. Nothing stays fixed but only gets worse. The Western limitation of not being able to understand the unseen renders it incapable of finding solutions to issues.

The reduction in crime throughout most of the country was nothing more than a temporary "fix" as unprecedented numbers of men were incarcerated. The prison population has doubled every eight years since 1976 when at 250,000 it was considered high by world standards. It leveled off for a while at two million in 2000 but is on the rise again. Reducing crime by putting men in prison does not address the cause, but adds to the cause by creating more broken homes. Crime is already on the increase and we will soon see another spurt in the number of incarcerations. Our society is unable to remedy any condition because it is unable to address its cause.

Community Board, PTA, and political club meetings change nothing. They are exercises in futility because they cannot deal with causes. To make change requires understanding the unseen and having the assertiveness to take action to bring about change. However, a society devoid of the masculine influence can only complain about conditions but it cannot change them, which is glaringly evident in media reporting. All news media report on what is wrong but have no concept on how to make anything right other than that the government should "do something."

Only men can make change–real men–those who understand the unseen and have the courage to take action.

Our nation is rapidly preparing the way for a tyrant for the reason that people can no longer resolve issues themselves.

A SENSE OF DIRECTION

I do ballroom dancing and enjoy dancing with inexperienced dancers as well as the experienced, as the former give me the opportunity to utilize my teaching ability. One of the mantras I recite to the learners is, "Back on the right and forward on the left." Not infrequently I have to stop to explain which is the right and left side. Women have difficulty relating to left and right, and also to North, South, East, and West. A woman's difficulty with direction is well known and the general attitude is "So what?" This essay will deal with the "So what."

First, let's look at how much of daily activity depends on an understanding of left and right. We turn to the left or to the right when we walk out the door of the buildings we reside in. We use even numbers on the left-hand pages of a book and odd numbers on the right-hand pages. Newspaper articles are designated left and right. On highways we pass in the left lane, and get off using the right lane. At the doctors office we explain of pain we have in our left ear or right knee, left foot or right eye. Athletic equipment is designed for left handed players and right handed players. There are left tackles, right guards, and left fielders. Boats and ships have left sides (port sides) and right sides (starboard sides). We look for things in the right side of a closet, and the left side of a storage bin. Refrigerators have left opening doors and right opening doors. The number of daily activities dependent on right and left designations borders on infinity.

North, South, East and West as designations influence all that we do. Streets are designated as running East and West, and North and South. We speak of the northbound lane of a highway or the southbound direction of an Avenue. We designate the side of mountains by the north face or south face. Subway and elevated lines are designated northbound or westbound. We speak of the southward flow of a river or the northward migration of ducks.

Women have difficulty with these designations of direction and others that will be mentioned because they cannot be seen. Not only are they not seeable—they do not exist. They are conceptual designations of men. Uptown and downtown fall into this classification, as do latitude and longitude. They do not exist. All of the movement of people throughout the world is dependent upon the conceptual designations of direction that men have created. That's something for both men and women to reflect upon.

Now let's take our thinking to a deeper level. Not only is all movement predicated on the unseen, but the rate of movement is also predicated on conceptual designations. There is no such thing as a minute or hour. They are arbitrary conceptual designations of the movement of the earth on its axis. Clocks are measures of conceptual thinking. The international dateline does not exist—neither do time zones.

Let's go still deeper, all that we manufacture is according to conceptual designations that do not exist. There is no such entity as an inch or millimeter, gallon or liter, foot, or yard. There is no such entity as a mile, a degree of movement on a circle. All direction, movement, and product of this world are predicated upon the unseen conceptual designations of men.

The dependence of all earthly activity on the conceptual thinking of men is an example of the functioning of the universal principle of gender. The masculine principle provides the environment and means for the feminine principle to bring life into this world and nurture it. The feminine principle produces everything in accordance with the unseen parameters of the masculine principle. Men deal with what is unseen in order for the seen to function.

We tend to take for granted the roads, highways, buildings, and products of society. We act as though they were always here. They weren't. They came into being as a result of the conceptual thinking of men. The attitude of "We don't need men anymore" is predicated on ignorance.

There is another area of male conceptual thinking that most significantly affects society.

All ethics and the morals that are derived from them come from the conceptual thinking of men. Virtue, chastity, honor, loyalty, and general rectitude come from the unseen standards of men. These standards came about as a result of the purpose of propagating and preserving the species. The enemy of our society had to disenfranchise men in order to brake down its moral structure. Removing the authority of men destroyed the family and the conceptual thinking within it that provided the standards for moral behavior. The enemy promotes the mantra "there is no right and wrong," and replaces ethical standards with laws that define legal and illegal.

Just as the conceptual thinking of men provided the environment for the activity of the physical world, the conceptual thinking of men also provides the spiritual, ethical and moral environment for the development of people.

All forms of government other than patriarchy are parasitic institutions that feed off the output of others. The huge federal economic stimulus prepared by the government is for it own survival and control. People can survive a recession and depression. They can survive famine. They have done so before. The government can't. It has grown too large. It needs tax revenue to survive. It must pump up the economy and in the process further destroy the environment.

We have a monster on our hands in Washington and in the nations of most of the world. That monster is destroying our lives. In the end it will destroy itself. That time is coming soon. For those who still think we don't need men any more, when the time of destruction is upon us the following words of Isaiah 4:1 will be fulfilled: "Then on that day seven women will take hold of one man and say, We will eat our own bread and wear our own clothes if only we may be called by your name."

Just as the conceptual thinking of God created the physical universe, the conceptual thinking of man created the material world in which we live.

A SENSE OF PURPOSE

As I review the societal landscape I see in its every activity a passiveness and lack of purpose. Life has become meaningless. Traditional practices, customs, and mores have become nothing more than empty rituals that gradually fade into oblivion.

Elders are no longer venerated; parents, grandparents, aunts, uncles, and other extended family members receive no special recognition and little respect, for they have been removed from any meaningful influence regarding the training and development of the young. Women no longer respect men and men no longer respect women, for their natural purpose has been suppressed.

Lacking purpose the only meaningful activity becomes self-indulgence. People who stand in the way of the attainment or expression of self-gratification become the enemy, regardless of their age or gender. When purposelessness engulfs a society its members become selfish and disrespectful; they develop a "dog eat dog" mentality, which requires an even stronger government to maintain order.

People seek escape from this un-natural existence in the form of entertainment, sports, substance abuse, gambling, and pornography. This entertainment no longer has meaning and has become a vehicle for the expression of more violent behavior and debauchery.

The purposeless society cannot endure; it will self-destruct. A purposeless society is an oxymoron. At best it is a temporary phenomenon.

Whether in the spiritual realm or material realm, purposelessness cannot endure. I will first address the concept of purposelessness in matters spiritual. Some of the new-age gurus preach of a society in which we will have made union with God and all conflict will have ceased. One sect established by such a guru holds that procreation will occur without sexual union and that desire and lust will have ceased to exist. I asked a member of such a sect what function the male would have in that society. She did not answer my question. She couldn't. If males provide the environment and means for females to bring forth life and nurture it, what possible role could man have in a society in which God has provided everything and there is no purpose? Man would not exist in such a society. Nor would woman, for she would not have anything to do. That society is purposeless. After union with God is attained, there no longer is a purpose or a reason for living in the material world.

The difficulty with Western thought in addressing this spiritual situation is that it can only comprehend it within the limited confines of the space-time paradigm, and spirituality does not fit in there. However, whatever understanding we have of spiritual perfection, we will no longer exist in the material world once it is attained, for the purpose of human existence will have been achieved.

Before proceeding, it is well to reflect that when the physical universe was created, gender came into being. Adam represented the masculine principle, which gave direction to the world in its process of fulfilling the divine purpose of propagating and preserving the species on its journey of spiritual growth.

Just as ultimate spiritual attainment removes the need for the male, the reverse is also true. As the male is removed from society it degenerates, implodes, and then ceases to exist. Whether we have spiritual attainment or materialistic degeneration, there is no place for the male in either activity.

In the former he evolved out of his position and in the latter he was removed from it.

No males—no purpose. No purpose—no society.

In the realm of human existence there are very few at the extremes—those that attain union with God or those that self destruct—but most of society tends to be somewhere in the middle of the spectrum, depending upon the virility and purpose of its males. Since males in Western society are impotent and lacking in direction and vitality, it is only natural that Western society is collapsing.

Whether or not one believes in the prophecy of the Mayans, Nostradamus, Edgar Casey, Revelations, the end times, or prophets such as Isaiah and Jeremiah, society has always been warned of the degeneration that takes place with the decline of the masculine influence. It happened, in ancient Israel, Greece, Rome, Spain, and now throughout the Western world. The words of the prophets are rarely heeded, and the lesson of history is that no one learns from it.

In order for a society to have purpose, or at least the ability to strive for goals, it must have virility. It must have the energy to fulfill its aspirations. Virility and the assertive action that it generates come form the masculine principle—from the male of the species.

No males—no virility. No virility—no action. No action—no accomplishment.

The neutering of the male in Western society has developed into a huge evil born of ignorance. It is leading to the demise of society and there is no "fixing" this demise. The health and well-being of men, women and children has never been in a worse state. Like an illness that must run its course before health is restored, the genderless thinking of Western society will wash away upon its collapse as the people face the issue of their very survival. That day is rapidly nearing.

We can stand by and do nothing, or we can start to prepare for that eventuality now. Those men and women who have supported Men's Action from the beginning might now consider moving forward to another level of activity. While we all have the demands of our daily existence on our available time, talent, and treasures, these demands are more apparent than real. Jesus told the man who would follow him but had a funeral to attend to first, to "Let the dead bury the dead." The meaning of that response is that most of the world is spiritually dead and let them care for one another. And another also said, Lord, I will follow thee; but let me first go bid them farewell, which are at home at my house. And Jesus said unto him, "No man, having put his hand to the plough, and looking back, is fit for the kingdom of God."

The time is at hand for those who are awake, to act, and for those who can see, to move. No institution will bring about change; it will be made by those who are awakening and emerging from the comatose state in which society is engulfed.

Society will soon be forced to re-establish its purpose and will require direction to attain that end. That purpose is the something for which to give up everything.

ATTACK ON REVEREND WRIGHT

The sermons of Rev. Jeremiah Wright created media frenzy during the primary campaign and they referred to his preaching as "hate speech." Political commentators and talk show hosts blanketed the airwaves with what they referred to as the dangerous and divisive talk of this African-American preacher.

On the other hand, the remarks of Rev. Parsley calling Islam "anti-Christ" and Mohammed "the mouthpiece of a conspiracy of spiritual evil;" and the remarks of Rev. Hagee referring to the Roman Catholic Church as "the great whore of Babylon" and a "cult," calling it an apostate church and stating that this false religious system will be totally devoured by the anti-Christ, received relatively little attention.

The comments of Reverend's Parsley and Hagee publicly condemned other religions in inflammatory and insensitive terms. Rev. Wright on the other hand didn't say that his religion was better than others or that other religions were wrong. He complained about the socio-economic condition of African-Americans, and the general attitude of America towards the rest of the world.

Rev.'s Parsley and Hagee did not receive much condemnation, for it is not unique for Western man to consider his views to be right and everybody else's to be wrong, but rather his normal modus operandi. As chief of Men's Action I receive comments from Catholics, Protestants, constitutionalists, libertarians, progressives, liberals, conservatives and people from other religious and secular institutions telling me that everyone else is wrong and if their belief system were implemented the nation would return to normalcy. Western man not only knows best, but also knows that he is best. The behavior then of Revs Parsley and Hagee might be a little irritating to the Western psyche, but not worthy of national condemnation.

Rev. Wright's comments on the other hand were inexcusable for he had the temerity to challenge the status quo of Western society. How dare he even think to do such a thing?

He forgot his place, which is subservience to the Western oligarchy. He had to be brought up short and reminded of who he is and isn't, and the media did that. And to support its efforts the media dug through its supply of black honkies to determine which ones could be called upon to make remarks condemning Rev. Wright.

Why must every African-American man speak up against some other African-American who makes inflammatory statements? Immediately after such an occurrence the media wants a response form the "black community." It does not call upon a response from the "white community" or "the

Jewish community" or "the Irish community" when one of their members makes inflammatory statements.

Total acceptance by the power structure of this nation requires a full embracement of Western thought and a rejection of one's cultural heritage. Reverend Wright understands this. He wanted others participating in the primary campaigns to become aware of it as well.

A TIME FOR RE-EVALUATION

As I skimmed through the 2009 Personal Finance Issue of US News & World Report and reflected on the disproportionate emphasis placed on matters material by the media, all levels of government, and our societal institutions I became motivated to write this essay on the true needs of people.

Contrary to the hype we receive from the various institutions of our society, the main interest and requirement of people does not evolve around money and things. When people pray to a higher power or go to fortunetellers, astrologers, or seers of various sorts they have three to four basic concerns. These are romance, health, and finance. Career also becomes an area of concern but it usually represents an extension of finance, and the finance serves as a means of material security. People want to experience health, love, and security. They have simple desires, which the extended family usually provided for except in cases of war and/or natural disaster; even then the feeling that "we are all in this together" provided a sense of security.

In patriarchal societies, women awoke in the morning in a safe and secure environment. They knew the men provided the protection they needed and the means for them to go about their business of nurturing. The men did the hunting, fishing, and farming necessary to provide the means for women to nurture, and provided for their safety as well, whether from predatory animals or incursions from a hostile tribe.

Women function well in a secure environment. It enables them to express their nurturing love in all its forms. It relaxes them and enables them to carry out their womanly functions with confidence, warmth, and joy. Women of patriarchal societies do not suffer from depression or commit suicide, and their bodies are relatively free of disease.

Men of patriarchal societies express joy and happiness as well. What greater joy is there to a man than living in a home filled with the aura of a caring loving woman?

Children raised in these societies have a sense of belongingness, which gives them self-confidence and a sense of responsibility to the larger social order. They also learn ethics and the natural moral behavior that ensues from these ethics.

Compare this picture of tranquility, security, togetherness, and harmony to the chaotic, insecure, lonely, and dissonant environment in which Western men and women live. Everything we do is based upon monetary decisions because family has been taken away from us. This focus on matters material cannot provide the security and warmth of the family and tribe.

Our government, media, and present society in general have a preoccupation with matters material and an almost complete disregard for the basic essentials of living and quality of life. Every human need becomes quantified and reduced to a purchasable material quantity. People no longer experience joy; instead they have temporary ecstasies provided by various forms of escapism.

None of the political candidates during the primaries or presidential campaigns focused on or even mentioned family. In all the hoopla about the activities of the President since the inauguration nothing has been mentioned about family. Even with health care—which is really sick care—the emphasis is on the amount of money available to cover sick costs, not wellness.

Western society has never experienced the level of illness that now runs rampant among all its social strata. Chronic illness now affects 25% of the population, diabetes occurs at younger and younger ages, obesity among children continues to increase, respiratory illness has become the norm, 300,000 new cases of breast cancer occur each year, and the number one health issue of the English speaking world is mental illness. And we call those areas of the world that still practice patriarchy, primitive. We have lost all perspective.

This perspective needs to be addressed, for it lies at the very core of the problems of Western society. A single entity created this universe and we generally refer to this entity as God although that word comes from the Europeans or Indo-Europeans. Mystics sometimes use the term "the great absolute" when referring to the single creator. Whichever way a person refers to this entity, it exists and is the cause behind all causes.

God created a perfect universe and gave instructions through various prophets on all continents since the beginning of time on how to live in this universe and be in harmony with it. There is a way to live in it. I will capitalize Way to denote it as a particular entity. The Tao means the Way. Jesus said, "I am the truth, the light, and the Way." When we live in the Way as best as we understand it and are capable of, we experience harmony, love, contentment and spiritual growth.

What we call primitive people lived in the Way. They had a balance between the material and spiritual world. They lived in harmony with nature, but knew all things came from the unseen spirit. They lived in harmony with each other. Patriarchy provides the structure for this harmony. It is the structure of the universe. It is an expression of the natural relationship and balance of gender. Western thought on the other hand is oblivious to the Way and therefore not in harmony with it.

Western thought only addresses the material world. It has little understanding of things unseen. It is feminine and materialistic. It is unbalanced. It does not acknowledge what it cannot see. It takes great pride in the scientific approach and use of empirical data, which one day I am confident will be regarded as an extremely unsuccessful attempt to access the truth.

Western thought has made a god of materialism. It believes it can create better food than God did, and that it can heal better than God can. Just look at the aberrations known as the American Medical Association and the American Pharmaceutical Association. It believes it creates abundance, not realizing that all abundance comes from God. It is not in harmony with anything and in disharmony with everything. Western thought has caused the pollution of the earth, and its inhabitants. All its political isms have failed.

Now as the economy continues to falter we hear continuous reports of the high priests in Washington calling upon the god of materialism to save us. The process of saving us will take away our every freedom.

It's time for a re-evaluation, either men rule or the State rules. The rule of the State leads to tyranny. The rule of men—which is patriarchy—leads to health, love, and security.

CHILDREN OF THE TYRANT MOTHER – PART ONE

An ordinary woman put in the un-natural situation of raising a family without a competent masculine presence becomes a tyrant. She tries to act as both the Sun and the Earth in the Solar System of her home, and because of her nature to do things in extremis she overcompensates and creates tyranny.

The tyrant mother rules with absolute authority, and if her authority is challenged, and her will not complied with, the consequences are devastating. One woman told me that she had been beaten with a hairbrush so badly as a child, that while she couldn't remember what she had done wrong, she did remember to never again get her mother upset. In explaining the tyrant mother to another woman, I said that every one wants to be close to mother because she is the source of love, but if one gets too close to the tyrant mother she is apt to lash out and claw you. The woman screamed these words, "That's exactly what my mother did when I got close. She clawed me."

Children naturally want to be close to mother and bask in her love, but children of the tyrant mother learn that it's safe to be near mom but dangerous to be close to her. When the children of the tyrant mother grow up, they find difficulty in being close to the opposite sex. In this age of promiscuity when having sex is done freely and openly, when virtue beyond the age of maturation is becoming a rarity, people are actually having less sex than they did 40 years ago. True sexual enjoyment requires intimacy; people cannot be intimate with a person they are unable to get close to.

Adults raised by a tyrant mother can't make personal commitments; instead they have "relationships." Relationships are like ice cream sodas; they're nice while they are going down but when they are finished, they are finished. Men and women need to strive for commitments, and the best commitment known to humankind is marriage, but how can one commit if one can't get close?

Another trait of adults raised by a tyrant mother is to marry their mothers, that is, they will seek out a mate who has the same traits as mom. Men and women alike do this. They subconsciously recreate the very environment that they claim to want to get away from. Since the environment of the tyrant mother was abusive, they create abusive environments. Spousal abuse has increased in Western society because both parties look to be abused. Instead of men and women blaming each other for their abusiveness they need to look in the mirror and realize the environments that they tend to create and/or seek.

The children of the tyrant mother were raised in a mixed-up environment. They became mixed-up and when they grew up they created mixed-up environments. The country is mixed-up. The whole Western culture is mixed-up. The number one debilitating illness of the American woman is depression, and the number one health issue in America is mental illness.

Have you seen enough? Don't you think it's time to make some changes and get society unconfused? Creating an environment in which family is central and men and women know their responsibilities and obligations to one another can bring about this change.

Next week the tyrant mother's effect on male conduct and effectiveness will be addressed.

CHILDREN OF THE TYRANT MOTHER – PART TWO

The route to adulthood begins at infancy and the home acts as the development center for this journey. The home environment teaches children how to relate to the opposite sex as they observe the relationship between their mother and father. It also enables their unique gender characteristics to unfold according to their age.

Boys raised in the environment of the tyrant mother have difficulty in relating to other boys because their manly traits received insufficient development. They have difficulty succeeding in any endeavor as adults. They can't quite grab the brass ring. They need mother's approval for what they do. They tend to associate more with women than men because the social environment of their home consisted primarily or completely of women. Also, they search for the nurturing and compassion that was denied them in their childhood. They have become losers. Most of them know that they are losers. It creates tremendous resentment within them knowing they have smarts but limited success.

Every boy wants to grow up to be a man—a real man—and doing so without a roll model to emulate causes confusion, self-doubt, insecurity, frustration, and anger. Most societies throughout the world recognized the importance of developing manhood and developed various forms of "rites of passage" to lead boys into manhood. At a certain age they were removed from the direct care of their mothers and placed under the supervision of men so that their manly characteristics could be developed and honed. Associating with men and boys developed a sense of comraderie; a sense of trust, cooperation, and understanding. They became secure in the company of their brothers and learned to work together in many endeavors.

Girls, even those raised by a tyrant mother, see how mother works, dresses, shops, and socializes. She sees other women come to the house and hears what they talk about. She learns about womanly things. What does a boy learn in a home without a father in it?

I became aware of the importance of a masculine environment not only as a child, but also as a father. My first child was a boy followed by two girls. One day my son said to me "You go to the office all day, you don't know what it's like to live in a house full of women." Eventually we had a fourth child, a boy, but when his older brother went off to college he said to me, "You go to the office all day, you don't know what it's like to live in a house full of women." They lived in a home where dad was at the dinner table every night, where dad played chess, touch football, and took them sleigh riding. Can you imagine what it is like for a boy to grow up in an environment where he never sees a man?

Boys raised in this environment are as crippled as a physically deformed person. Their psyches have become deformed. They become sensitive to female criticism because moms were the only authority in their homes and to be criticized by her was dangerous. Women do not like to marry a man with a strong mother behind him. They want their own real man. They are finding the pickings very slim, because tyrant mothers can't raise real men and society no longer places an emphasis on raising them.

CHOOSING YOUR SIDE

My essay *Elimination of the Family* received an excellent response world wide, as did others that recently addressed family and the well-being of society. This response indicates an increased awareness on the part of a small but growing minority of people worldwide that something fundamental has gone wrong with society. They have begun to realize that conditions will force them to choose sides between those who believe in surrendering their lives to the government, and those who prefer to maintain responsibility for their own well-being.

With each passing day Western governments cement their control over the people by enacting onerous laws that make people completely dependent upon an omnipotent and omnipresent ruling structure.

One of the many devious methods they use to maintain their hold on the populous is to pass laws that seemingly increase justice and fairness to all, such as anti-discrimination laws against race, ethnicity, religion, sexual orientation, and gender. However, the effect of these laws eliminates an ethical basis for society to live by; it also punishes those who promote such values. A Canadian Roman Catholic Priest who was accused of committing a hate crime against homosexuals because he quoted form the Bible represents the invasive and constricting effects of government laws. I don't think it is far-fetched that I could be accused of committing a hate crime because I espouse patriarchy and the rule of men. I could probably be accused of a hate crime because I have the temerity to preach that men and women are different.

We have reached the point where our government considers itself the establisher and administrator of all societal values. I challenge that position. However, before pursuing that challenge I want to make my personal position regarding humankind clear.

I consider every man and woman to be my brother and my sister, my son and my daughter. Every is an all-inclusive word and does not require further elaboration. However, loving every human being does not mean that every human being can fit into any role in society. Equal worth does not mean sameness. Every organization has its requirements and its standards. All drug users are my brothers and sisters, but I would not want them in key positions in my organization.

Walt Whitman expressed this dichotomy of thought in his *Song of the Open Road*, in which he states...."The diseased, the illiterate person are not denied; the beggar, the drunkard, the escaped youth; are all included in society." But then later he states, .."None may come to the trial till he or she bring courage and health, come not here if you have already spent the best of yourself, no diseased person, no rum-drinker or venereal taint is permitted here." His words may seem to be a contradiction; he accepts all of life, but those who travel with him must be of a certain standard.

Jesus administered to the wine imbibers and all manner of social outcasts and misfits; however, his inner circle of disciples was made up of intelligent and spiritually tutored men. He had his standards.

So we as a society must have our standards. While none are denied, only those who understand and adhere to the standards will lead.

The issue that I take with the government and its proponents is that they have no standards; they offer rights, which only they can bestow. They have made unlimited self-gratification the objective of society. I believe Ethical values need to be re-established, and that the propagation and preservation of the species serves as a primary determinant of ethics.

It is on this point that you will choose your side. You either work for the propagation and preservation of the species in a structure administered by men called patriarchy, or you support laws that preserve the right to equal self-indulgence for all and is administered by the state. All political systems, which by the way came from Europe, have failed. Therefore, choosing government is choosing failure. Patriarchy on the other hand, is the natural structure of the universe and the natural structure for humankind. It has been with us since the beginning of time.

If you choose to work for the propagation and preservation of the species you will associate with those people whose values and temperament support that objective; they will be drawn to you and you to them.

This is the issue.

Men make ethics that govern moral behavior; men set standards; men determine what is good for the preservation of the race. Women bring life into this world and nurture it. That's our partnership. The value to society depends upon people's ability to contribute to the propagation and preservations of the species including its spiritual growth.

The time has come to make your choice; will you help to draw the battle line, or will you submit to the growing tyranny?

CUBISTS

A newborn child begins his cubist journey through Western society when he is placed in a cube called a nursery. The child leaves the hospital in another cube called a car, which brings him to another cube called a home, in which he will be brought to another cube called the baby's room.

Everything that the child does and sees occurs in a cube. His parents take him out in the family cube and bring him to other cubes such as stores, offices, and restaurants. When he starts school he is picked up in a cube called the school bus that drops him off at a large cube called a school, where he goes into a smaller cube called the classroom. As he gets older he travels in other cubes such as subways and railroads to go to other cubes such theaters and museums. All his learning takes place in a cube, and when he finishes school and seeks a job it might be in a big cube called an office building or a factory, where he might work in a little cube called a cubicle. If the child grew up in the city or even suburbia, he had very little exposure to life outside the cube. On those occasions when he spent time outside of the cube he probably couldn't assimilate what he saw because his senses hadn't been developed to respond to non-cube influences.

The cubist environment has a devastating effect on the functioning of those raised in it, and essentially conforms to the results of environmental tests done on cats to determine the effect of early environment on their sensory perception. Kittens in one group were raised in a room that only had vertical objects and vertical lines on the walls; kittens in the second group were raised in a room that only had horizontal objects and horizontal lines on the walls, and kittens in the third group were raised in a room with no objects and with walls painted a solid color. When let out of their rooms upon reaching adulthood the first group bumped into horizontal objects, the second group into vertical objects, and the third group bumped into everything because they could discern nothing. These tests explain why people who were born blind and had their sight restored cannot distinguish a circle from a square without holding it in their hands; they have not yet developed the visual ability to distinguish between the two shapes.

The effects of early environment upon cognitive ability became apparent in the ghetto area of Chicago during the civil rights movement of the 60's where most first grade American born black children could not comprehend a story about a train trip to visit grandma who lived in a house by the lake. The children were considered to have low IQ's until some thinking person made the observation that these children had never seen a lake, never ridden on the railroad, and grandma was someone who lived with them or did not exist. The first grade story book had no relevancy to the environment in which the children had been raised, as the circle and the square had no relevancy to the visual acuity of the blind man, and as the cats could not discern what they had not been exposed to as kittens.

The effect of early environment on ability to function in later environments is understood by behaviorists and sociologists, but there seems to be a complete lack of understanding about the difference between a natural and an un-natural environment, and for good reason. Behaviorists and sociologists along with their colleagues in other disciplines are cubists; they have been removed from what is natural and their thinking is conditioned by what they learned inside the cube.

Cubes do not occur in nature, nor do straight lines and flat surfaces, nor do single colors and monochromatic colors, nor do monotones. Forests are not green, they have vegetation composed of all shades of green that grow on bushes or trees that have all shades of brown up to black, and have stones and rocks around them with a full spectrum of coloring. Light intensity continually varies due to the movement of the trees in the wind and the clouds in the sky, and the sounds of the forest come in various pitches and intensities. The forest floor has irregular shapes and densities. The forest is a pulsating, gender motivated, natural environment; God created it.

The environment of the cube differs completely from that of the forest. Cubes are usually painted in one solid color, fluorescent bulbs provide constant intensity lighting, its floor is flat and of one density; and the sounds whether coming form the air conditioner, fan, or radio tend to be of one pitch or of extreme dissonance. A cube is a static, genderless, unnatural environment; man created it.

Cubists know a lot about the workings of cars, planes, stereos, computers, cell phones, and videos; however, they do not know the name of the tree that brushes against the window of their cube, or the bird that sits in its branches. Cubists think they are smart, modern, and enlightened. They view people who live outside of the cube as primitive and ignorant. In keeping with his Western, feminine, materialistic training the cubist believes his way of life to be right and wants to educate others to be like him. It does not dawn upon him that those who dwell outside the cube live a more natural and spiritual life that keeps them healthier and happier. The cubist's cognitive ability has not been developed to see that.

DEPRESSION OF WESTERN WOMEN

There are many indications of the deterioration of the well-being of American women such as high rates of breast cancer, obesity, diabetes, and migraine headaches; however, the most telling characteristic of the decline of the well-being of women is their deteriorating mental state.

The single most significant measure of the state of women in our society and in Western nations in general is their high level of mental illness. The number one debilitating illness of the American woman is depression with upwards of 10 million suffering from this ailment. Three million girls also suffer from depression. Depression has become so widespread that a major criterion in selecting colleges is the competence of their psychiatric staff, and a major criterion in evaluating college applicants is their psychological health.

Antidepressant drugs such as Prozac, Luvox, Paxil, and Zoloft have become household names. Children today probably know these names as well as they do Bayer Aspirin. Knowing the name of these drugs they then think it normal for their mothers to be depressed, and it is normal—not natural—but normal in our society.

Drugs—all drugs—produce dangerous side effects in the body further deteriorating the health of the women who use them. Studies are being made (Western thought loves studies because it knows nothing) on the causative factors of depression among women.

The following is an excerpt from one such study made by a woman who is attempting to blame society for the depression of woman:

> *The increase in the number of women diagnosed with depression, sometimes when symptoms are minimal, has come on the heels of an approach to treatment known as "biological psychiatry." In this approach, rather than looking for social, cultural and economic and life stage factors that might be making a woman depressed or anxious, doctors are taught, and patients have come to believe, that the cause of symptoms is biological. This makes it seem logical that a drug is needed and appropriate.*

This conclusion is partially right; however, the writer didn't explore the social cultural and life stage factors that needed to be addressed. An analysis of American women who suffer from depression shows that a disproportionate share consists of those who are white, middle class and single, especially single women with children. This is a social, cultural, and economic grouping whose constituents make up the primary membership of feminism and "women's liberation" thought. They have become so liberated that they are going out of their minds. They have no concept of family.

Feminists are the one's who've been on psychiatrist's couches, who belong to support groups, who are on prescription medicines, and who carry anti-depressants and painkillers in their purses.

Western women throughout the world suffer from a disproportionate amount of depression; in Europe depressed women comprise close to 15% of the female population. That's a disaster. Yet they make up a significant part of the electorate that elects women to political power. Isn't that amazing? The depressed ones—the nutsey byes of society—wield the political power that affects our lives and in particular laws affecting marriage and the care of our children. It is only in the Western nations that women have risen to political power without some relationship to their husbands and their fathers. It is these Western nations that have the highest rates of female depression.

Those nations where women do not rise to political power have the lowest rates of depression. They also have lower rates of breast and cervical cancer, suicide, divorce, and adultery. They also have children less likely to become unwed mothers, join street gangs and end up in prison. They do not live un-natural "liberated" lives; they live in happy and secure families. The media treats those women as unenlightened, third world people. This is a true travesty of values.

The liberated independent Western woman in the aggregate is the unhappiest woman in the world and she produces children who are likewise. The reason that these women are depressed is because they have been removed from their natural environment and been deprived of their natural function—to bring life into this world and to nurture it. They can only do this if men provide the secure environment and means for this nurturing to take place.

This secure environment is known as the patriarchal structure and it produces family. Men make patriarchy. No men—No patriarchy. No patriarchy—No family. No family—No security. No security produces depressed women.

The Western woman is suffering from depression and there is no government program that will improve her lot.

Surveillance cameras and security guards do not alleviate the problem. There is only one species of life that protects, cares for, and loves women. It's called man.

DESTRUCTION OF ETHICS IN WESTERN SOCIETY

The terms morals and ethics tend to be used interchangeably, but the former depends upon the latter. Ethical standards determine moral conduct. Moral behavior is the manifestation and personalization of unseen ethics. We have here another example of the principle of gender at work. The unseen assertive masculine influence acts upon the seen receptive feminine entity. By eliminating the authority of men Western society has destroyed its ethical base, rendered its organizations impotent, and fallen into a state of increasing chaos.

Ethical men form organizations to achieve certain purposes. Ethics and discipline provided by leadership serve as the glue that holds organizations together. Rather than address what constitutes capable leadership, effective discipline, and a high level of ethics, this essay will focus on the fundamental reality that men build organizations and that ethics play an intrinsic role in their operation.

Men build organizations when they realize the need to cooperate with others in order to accomplish an objective, whatever that objective might be. They build teams, armies, associations, crews, brotherhoods, hunt parties, and families.

The fundamental objective of human association focuses on the propagation and preservation of the species, which is accomplished through the family—the first organization. All the other organizations that men developed centered on sustaining the family and then the extended family or tribe. Men create defense groups to protect the tribe, they develop hunt parties to obtain food, and they develop mating rituals to propagate the tribe.

In order to determine who can best fill certain positions in the organization men observe and judge each other's abilities. They do this in a spirit of cooperation for they all understand that they need the most capable man in each position to best ensure their own survival and that of the tribe. It is important to know who can run the fastest, yell the loudest, lift the most and last the longest. Sports were developed as a means of evaluating the ability of men to serve the needs of the tribe. It also gave women the opportunity to assess potential mates.

Discipline develops the reliability, dependability, and constancy of the group. Rules of conduct develop in order to maintain the stability, viability and growth of the organization. Ethics provide the base for these rules; a major ethic of non-western society is to share. Ownership is minimal in non-industrial societies; people share what they have with others, and others act to assist those in need such as helping to repair another's hut, wigwam, or house. Widows and orphans are cared for and no one is left out. Ethics developed in non-industrial societies focused on the training of the young, the care of the aged, and the safety and well-being of all.

All organizations at their inception are patriarchal. As the masculine influence is reduced in an organization its vitality declines and its ethics disappear, a condition spreading rapidly throughout Western society, and as we will see, by design.

Men create organizations that are inclusive. Women develop cliques, which are exclusive. Cliques develop among pre-kindergarteners, continue through grade school, high school, college, the work place, the community, and in any organization that contains a group of women. The formal name for clique is aristocracy. Feminine oriented Western society always creates aristocracies. Apartheid, segregation, and economic classes are manifestations of the Western feminine aristocratic nature.

Certain advantages do accrue from aristocracies, but they will not be explored here other than to say they cannot be established, but instead evolve out of an existing structure. All organizations at their inception are patriarchal.

Some will point to the National Organization of Women as being a large powerful international and independently operated women's organization. It is nothing of the sort. Why aren't they located in Chicago to be equally accessible to women from all over the country? Why not in California with its large liberal base? Why not sunny Florida? NOW locates its headquarters in Washington DC so that it members can maintain a tight grip on the weak-kneed wimps who run our country. How do they maintain their grip in Washington and in all the state legislatures where they have lobbyists to promote their cause? Where do they get the millions upon millions of dollars necessary to carry out their activities? Who gives them direction? What is their dogma? What have they contributed to society?

Feminists are nothing more than the dupes of an international elite bent on controlling the world. This elite provides the thought, money, and direction for feminist activities.

At the 4th International Conference on Women held in Beijing, China in 1995 the feminist objective was to eliminate gender and religion from all societal structures. The powers behind the feminists knew that if gender were eliminated, dynamic virile organizations could not be formed or maintained. The elimination of gender also weakens religion—a necessary component of ethics.

The results have been spectacularly successful for those behind the intent of that conference. There are no virile organizations remaining in America. Western Christendom is completely impotent and has been reduced to conducting empty rituals. All standards and values have declined. The adage "do not judge" has been taken out of the context of its spiritual meaning and has become a mantra that condones all behavior.

Male organizations have been eliminated under the guise of providing equal opportunity for women. Opportunity for what? To make money? Making money for its own sake is whoredom, which is a subject for another essay. The true purpose behind the elimination of men's organizations is to destroy the source of ethics. Without ethics all society degenerates. As the degeneration continues people cry out for help. With the elimination of viable men's groups the government looms as the only alternative source of help. We beg it for help. We beg it to protect us. We beg it to feed for us. We beg it to be our masters and for us to become its slaves. We are getting what we begged for.

As I write this article the realization of the magnitude of the destruction of society caused by its lack of understanding of the difference between those who have testes and those who have teats, overwhelms me. My brothers and sisters willingly walk into the lair of the demon unable to do otherwise.

The elimination of tyranny and the slavery it produces, can only occur when men of good will band together and act. Good will is a function of ethics, and the banding together is organization. Action comes from the WILL of organization members. No ethics—no organization. No organization—no constructive action. No constructive action—no escape from tyranny.

The only species of life designed to provide for the well-being of women and children is called man. Men provide well-being through ethical based and active organizations.

You can learn more about such an organization on www.mensaction.net.

DEVASTATION OF WESTERN WOMEN

Feminists have taken deliberate overt action to destroy the value structure of women in order to motivate married women to leave the home and single women to avoid the institution of marriage altogether.

Removing women from the marriage environment served as the vehicle for fulfilling the ostensible purpose of giving them independence and equity in society, including entry into all organizations that men made. This goal of parity masks a more sinister objective the effects of which have produced devastation among the ranks of women and a deleterious effect on society. The time has come for the unmasking of this sinister objective and then for a return to a naturally functioning society that cares for women and children. However, we first need to determine the reason for the creation of the institution of marriage along with related institutions that have become subject to attack.

The mating ritual of most all life results in the forming of a reproductive structure that supports the propagation and preservation of the species. Among humans that structure has become known as the institution of marriage and it existed long before the enactment of any governmental laws to sanction it.

Women bring life into this world and provide its nurturing through their natural attributes such as adaptability, responsiveness, and visual thinking. They can only fulfill their function in an environment that provides them with safety and security. The highest order of activity for men consists of providing the safety, and security that enables women to bring life into this world and nurture it.

The attributes of constancy, assertiveness, and conceptual thinking enable men to provide the environment conducive to nurturing through the organizations they create and by the ethics they develop. These organizations and their ethical foundations interweave into the development of cultural mores and religious rites that in one way or another relate to the propagation and preservation of humankind.

By removing male authority from the home the institution of marriage has crumbled resulting in high divorce rates, low marriage rates, and an explosive rate in unwed motherhood, which in 1950 hovered between 1% and 2% in America and has now reached close to 40%. The same increases have taken place on the European continent with unwed motherhood near 40% in England and 50% in France. The negative effect on the children of unwed mothers has far reaching consequences such as high drop out rates in school, increased likelihood of a relationship with the law and prisons, and the prospect of their daughters becoming unwed mothers as well. The effect on women of unwed motherhood and the efforts of the government to deal with it such as promoting birth control pills, condom usage, and abortion, has been devastating.

By destroying the institution of marriage women also lost their prestigious positions of good wives and mothers. All through history women who were raised to be good wives and mothers were sought after by men and revered by society. Women dressed and acted in a manner to be attractive to men and men dressed and acted in a manner to be worthy of having such a woman as his wife. Women lived in a "win-win" situation. No man could compete with a good wife and mother.

Women no longer have that prestigious position because they are no longer considered to be women. They have become persons instead. What is a person? What does a person do? How many ethnic groups have the word person in their vocabulary? Most of the world refers to its members as men and women. The Bible states that God created men and women. What part does a person play in the propagation and preservation of the species? Persons do not desire to become attractive to other persons; their dress and behavior reflects this lack of desire. The attire of men has become slovenly and that of women runs the gamut between dressing as a street person or a porn queen.

Women do not naturally feel any joy in being a person. The motivator of a woman's life is the ME aspect of her personality. The ME factor motivated a woman to become a good wife and mother. How does personhood satisfy the ME aspect of a woman? It doesn't; other criteria have been developed in an attempt to satisfy the ME aspect of the female psyche.

Girls are now taught to go to school, get and education, get a job, and make some money, so that they can be somebody. A big step towards becoming that somebody requires obtaining a piece of paper that indicates completion of certain areas of rote learning. That paper will serve as a pass to do what ex-men do (relieved of their former authority they have now become jackasses on the treadmill of production). These pieces of paper that women receive entitle them to join the jackasses on the treadmill. Regardless of the name given to that piece of paper—whether degree or license—it falls under the greater heading of what I call "she-ass certificates." These "she-ass certificates" serve as the primary criteria for hiring and admittance to the treadmill of production.

Part of the requirement of utilizing these she-ass certificates necessitates turning over to the state the responsibility of all aspects of child rearing. The state now sets the requirements for health, education, discipline, food, and dress for children. If these requirements are not met the child is taken from the custody of the she-ass certificate holder and given to other persons that the government deems qualified to raise children according to their standards.

Since men have been removed from all levels of authority and their organizations and institutions weakened the meaning and purpose of life has become obscured and ethics have disappeared as is evidenced by the removing restrictive laws on incest in Europe. Three European Union nations -- France, Spain and Portugal -- do not prosecute consenting adults for incest, and Romania is considering following suit. It will soon become the norm in the Western world to conduct sexual relations between mother and son, father and daughter and brother and sister. Livestock raisers know that this type of inbreeding weakens the herd, but in a genderless person society apparently this fact has little significance.

The establishment of a genderless society had nothing to do with the liberation and benefit of women; on the contrary, women have lost out on all fronts. They are being used and abused and consequently suffer from mental, physical and spiritual deterioration. Feminists served as dupes of a sinister conspiracy to have government control all of society. By neutering the male all institutions designed for the propagation and preservation of the species were weakened or destroyed, which enabled the government to step in and take over the administration and control of every aspect of societal activity and in the process has reduced men and women (or persons) to the role of jackasses and she-asses on the treadmill of production.

Government can provide substitute authority but it cannot provide substitute ethics, which can only come from men; consequently we are witnessing the complete moral decline of the Western, genderless, person oriented world, which is responsible for the devastation of its women.

ELIMINATION OF THE FAMILY

All groupings have an authority from which all direction emanates.

This truism applies to insects, fish, flora, mammals, and humans. It also applies to what we refer to as inanimate life such as planets and suns, and atoms and molecules.

All organizations such as clubs, teams, businesses, and religions have their authoritative heads. Community, fraternal, educational, governmental, and social groupings have their top authority. This authority might be elected, appointed, inherited or just agreed upon by its members and might be called, chief, chairman, president, manger, coach, captain, priest, rabbi, imam, pastor, or leader.

Every grouping contains a directing influence that guides it to its collective destination, or at the least, helps to preserve it. This structure and activity is a manifestation of the universal principle of gender; the assertive masculine influence providing the environment and direction for the receptive feminine entity. No action can take place in the universe without the interplay of gender.

The largest celestial groupings known as galaxies consist of millions of stars and planets held together gravitationally and moving collectively in the same direction through space. A subcomponent of a galaxy is a solar system. All the planets within the system move in the same direction as the Sun and are completely under its control. In the atom the electrons—the feminine principle of electricity—rotate about the nucleus. From the smallest atom to the largest galaxy the unseen power of the masculine principle provides the direction and structure of the grouping.

Every flock of chickens has a rooster at its head and every gaggle of geese has its gander. Every herd of deer and cows has it stag or bull to lead it. The male lion heads the pride. No group can survive without a head, leader, or final authority. We all know this.

Why then has society removed the head of the family?

There are those who say that the family has become obsolete. It hasn't become obsolete; it has been eliminated. Removing the man as its head had the same effect as removing the Sun from the solar system, it would collapse and cease to be. We have removed the man, family has collapsed and ceased be. We have produced a generation that believes in government as the provider of its well-being. It has no choice; the family has been eliminated.

The elimination of the family has been accepted on a national basis. None of the presidential candidates mentioned family in presidential primaries, neither party mentioned it in their platforms, and the inaugural address made no mention of it. The psyche of the nation no longer focuses on family.

The philosophy of Plato—which called for the turning over of children to the state—has now become a reality. Mothers actually brag at how early they have entered their children into pre school programs. We insist on schools feeding our children instead of parents preparing their lunches at home. We have eliminated home taught ethics and the resultant moral behavior and substituted it with after school indoctrination of legal and illegal. We have become a nation of automaton workers and programmed consumers—jackasses and she-asses on the treadmill of production—while the government controls every aspect of our lives. We brag about that condition as though it were an enlightened way of life.

We ask the government to provide health care, economic well-being, and educational opportunities. Government has been made into a God upon whom our very lives depend. Accordingly, we worship this God and the priesthood that carries out its instructions.

Let me be clear—the family has been destroyed by ignorance of the universal principle of gender. No amount of legislation or government tinkering can re-establish the family as a viable grouping. Government destroyed the family; how could it possibly re-establish it? Re-establishment of the family can only occur by working outside of the government, outside of the system, and by realizing that a family must have its head and that head is man.

The function of all life is to propagate and preserve itself so that we can grow spiritually. The masculine principle provides the environment and the means for the feminine principle to bring forth life and nurture it. It does this through organization, structure, and ethics. The traits and characteristics of each gender become the attributes that enable the propagation and preservation of the species.

Family is an organization. Like all organizations it requires leadership. The male is designed for that role. There is no substitute.

If you understand that family is the only viable structure for human kind and want to do something about it, then become involved in Men's Action. We have a message. We understand the universal principle of gender and know that society can only function properly with the natural interplay of gender called patriarchy. Patriarchy is family.

Our mission statement: To Foster a Natural Way of Life for Humankind, is patriarchy.

Men make families. The state makes slaves. You have two options and one choice. Which will it be, the rule of men or the rule of the state?

FATHERHOOD IS NOT A JOB DESCRIPTION

Fathers received attention this week as a result of the President's message appearing in Parade on Father's Day entitled "We Need Fathers to Step Up." There was also some media attention given to the letter he wrote to his daughters two days before his inauguration. Whether they were his words or those of his writers they conveyed the impression that he is pro-family and understanding of the responsibilities of fatherhood.

His message might have had more credence if he had not chosen as his vice-president the sponsor of the Violence Against Women Act, the enforcement of which has resulted in the increased incarceration of men, removal of fathers from their children, and the further breakdown of the American family. It might have had more credence if he did not wax eloquent calling it a "wonderful day" at the signing of his first piece of legislation, the Lilly Ledbetter Fair Play Act of 2009.

If you question what these two pieces of legislation have to do in addressing the issue of fatherhood, they were both designed to usurp the authority of men.

Regarding the Lilly Ledbetter Act, its focus was on equal pay for women. I am a former owner of a small manufacturing business and during the 23 years that I ran it I paid people according to the contribution made to the company. I paid them fairly, a policy that at least partially accounts for never having been approached to unionize. I made the decision as to what constituted fair pay, not some bureaucrat in Washington, the state legislature, or city administration. Since the enactment of the Lilly Ledbetter Act the judge of fair pay will come from a source outside of the business owner and will further erode the authority of men to conduct business according to their own judgment and ethics.

A fundamental of all organizational operation requires that authority be given to those with responsibility. The President has a Cabinet and every member in it has the authority to run his department as he sees fit, striving to attain the goals set by the president. There is no organizational structure that can long endure without giving those with responsibility the authority to do what is necessary to fulfill that responsibility.

Put another way, every organization has its head, whether a business structure or societal grouping. Every corporation has a president and every herd has its buck. There can only be one leader and that leader must have authority.

The government has destroyed the family structure by usurping the authority of the male; therefore, it no longer refers to the family in domestic relations but instead speaks of fatherhood and motherhood, reducing both activities to job descriptions prepared by the government. In keeping with the ever-growing emphasis on a genderless society, motherhood and fatherhood are being increasingly referred to as parenthood.

Fatherhood is not a job description; it is a function of developed manhood. The process of fatherhood begins with the development of men—real men—who understand that their primary function in life consists of providing the environment and means for women to bring forth life and nurture it.

When a real man decides to start a family he seeks a woman who has what he considers the desirable attributes to be the mother of his children. When he finds her and she in turn decides he has the attributes that will provide the environment and means that will enable her to nurture a family, they agree to mate. This mating agreement is formalized to the community in a ritual known as the rite of marriage, in which the man and woman are making a commitment to the community to bring forth life and raise it to the best of their ability. After this ritual they become known as husband and wife.

As husband and wife they begin preparing the nest in which they will raise their offspring. This nest or home will vary according to the social mores and customs of the husband and wife, and their particular economic strength. After the first child is born, they will become known as mothers and fathers. The progression of fatherhood is from man, to husband, to father. Real men make good husbands, and good husbands make devoted fathers.

This progression applies to women as well. A good woman makes a good wife makes a good mother. To think that it is possible to be a good mother without providing the child an environment containing a good father represents fallacious thinking at best. Ninety percent to 95% of our prisoners come from homes in which there were no fathers.

Government has removed all male authority; consequently there can be no family. There will eventually be no free enterprise either. The elimination of family has left society without a cohesive bond. Government attempts to administrate the relationship between men and women, mothers and fathers, and parents has resulted in a national angst caused by a lack of belongingness. Fatherhood in that arrangement becomes a job description such as carpenter or stockbroker. It becomes devoid of any obligation towards the propagation and preservation of the species including its spiritual development. Government has no spirituality; nor does it possess any ethics. It has replaced moral and immoral with legal and illegal. It is this thinking that enables taking a pro-abortion stance while publicly proclaiming support for fathers. The tie in with abortion-on-demand and the usurpation of male authority is either not recognized or not acknowledged.

Only men make families. Only men make any organization. The disempowering of males leads to the collapse of society. Women cannot be empowered vis-à-vis men any more than the earth can be empowered vis-à-vis the Sun. The Sun can be disempowered, that is, its energy can be diminished, it can lose its pull of gravity, and eventually as it becomes more like the earth the Solar system will fall apart. The earth on the other hand, cannot in any way become more like the Sun. So too in the relationship between men and women, men can be castrated, neutered, and emasculated, but the woman can never become empowered. A sex change will never occur in which the new male can be empowered with virility.

What this society calls the empowerment of women is the disempowerment of men by law and the usurpation of this power by the government reducing men and women to automaton workers and programmed consumers.

Having reduced fatherhood to a job description our youth will continue to do poorly in schools, teen pregnancies will remain the norm, ADD will increase among our youth, the rate of marriage will decrease, and the rate of divorce will increase. These conditions do not bode well for women since 10% of them already suffer from debilitating depression in America, as do 15% of European women.

Until such time as men retake their authority and re-establish themselves as head of the family government will continue to extend its tentacles into every facet of society reducing it to a huge ward of the government.

Fatherhood is not a job description. Don't accept it as such.

GENDER AND RELIGION

Newsweek's cover story for the April 13th issue entitled *The End of Christian America,* provided me with the motivation to address religion from a gender perspective—something I had intended to do for a long time. However, as I started writing this essay I realized that the Newsweek story came out during Easter Week. For an international news magazine to promote its lead story with the words The Decline And Fall Of Christian America written in large red letters on a black background of its cover during the holiest week of the Christian calendar was an "in your face" attitude of disrespect by the media.

This blatant public disregard of the feelings of the Christian community in essence proved the point of the publishers. What would or could the Christian community do about it? The article—like most that appear in major news magazines—covered both ends of the spectrum that they addressed in an effort to show fairness; however, the headline and title of the article delivered the message. This attack on religion by the media and the powers that control it is part of a planned effort to destroy religious influence in society. I have addressed this issue in other articles and will now continue with the original intent of this essay.

The Newsweek article expressed little understanding of the reasons for the waning of religious and non-governmental institutional influence. I will comment on both of these conditions, as one bears upon the other. The reader familiar with my message will see gender woven into many paragraphs.

Religions are feminine in nature and reflect the materialization and personalization of unseen spiritual truths. They also represent the understanding by the local populace of matters spiritual and its resultant method of worship, which accounts for the same religion having dissimilar practices in different parts of the world.

Religions represent the common understanding and interpretation of the unseen by a particular group of people. This understanding might be based on scripture and/or from the sayings of prophets and other categories of "holy men."

Changes in religion are not changes of the truth, which cannot be changed, but in the understanding of the truth, which continually changes. The Christian religion had its first change in understanding when Eastern Orthodoxy broke away from Roman Catholicism. The next major change came with the Protestant Reformation, which ushered in several different denominations. Every major religion has its various offshoots or sects, which represent groups of people with a similar understanding of matters spiritual.

The practice of considering religions as synonymous with scripture reflects erroneous thinking. Differences—sometimes vast—exist between religious practices and their sources. Christianity in particular was formed by a feminine materialistic culture more than three hundred years after the death of Jesus and spearheaded into acceptance by a Roman emperor who worshiped the Sun god. All liturgical Christian denominations derive their liturgy from the ritualistic worship of the Sun god for which there is no basis in the vast spiritual compendium called the Bible. Significant remnants of the European culture in the Middle East during the time of Mohammed influenced the interpretation of the Koran and the practice of Islam.

The spiritual understanding of a religion and its ability to communicate it varies inversely with the authority required to maintain its influence. The more that a religion can teach, guide, and train its members, the less authority it needs. Religions that preach blind obedience to dogma require complete authority over their members.

As spiritual awareness of people grows they become less inclined to have religious authority imposed upon them and instead seek to increase their understanding of themselves and their world. On the other hand, people want a certainty and constancy of matters spiritual. They want religion to be authoritative without it becoming authoritarian. The European influenced religions have never offered much enlightenment to its members and now that they have lost their authority they are rapidly losing their influence in society.

The societal structure in which religions function also affects their influence upon that society. Religions little influenced by European thinking offer great enlightenment and seemingly hold little authority over their members even though vast populations operate in accordance with their teachings. These religions function in a patriarchal society, which by definition has authority. There is respect for the father, the mother, for elders, for ancestors, for everybody and everything. Religions do not need added authority in patriarchal societies. Society IS the authority and all activity centers about the well-being of the family—the core of all patriarchal societies. You can be sure that the gurus, swamis, and enlightened masters that now offer their enlightenment to the Western world did not do anything without the blessings of their fathers.

We do not have such a society in America or the Western world, hence the Newsweek cover article indicating the decline of Christian influence in America. Indeed it has declined, and for two significant reasons. It has failed to provide enlightenment of the true meaning of Biblical scripture and instead relied upon fear to maintain its hold on the people. Secondly, all institutional authority in America and the Western world has collapsed.

While all religions are feminine in nature, the structure in which they function is masculine. Men build organizations; an inherent trait that manifests early in childhood. Boys form clubs, teams, and gangs. They know the abilities of each of their members: who can yell the loudest, run the fastest, lift the most, spit the furthest and even pee the furthest. These traits are utilized when it comes time to accomplishing a task that requires the participation of all members. As the boys grow into manhood the purpose of forming groupings becomes manifest—to provide the environment and means for women to bring forth life and nurture it. Whether to form a sailing crew, hunting party, or defensive detachment, the objective remains the same: the care of women and children.

Now that Western governments have done a comprehensive and thorough job of legislating away all male authority, all organizations and institutions have lost their virility. Civilizations decline when virility declines. All institutions and organizations in the Western culture have declined in influence and now rely on government support and involvement in one way or another as it attempts to become the substitute male.

The institution of the Church requires the same masculine leadership as any other institution. Its failure to provide this leadership is self-imposed. Focus on the Family for example, in its three decades of existence never extolled the virtues of manhood. It believes it can build families with fathers but without men. It does not understand the patriarchal structure of the universe. This thinking permeates all of Western society and consequently the influence of all its institutions are declining including its religious institutions.

As humankind continues its quest for spiritual meaning, the trend to meditation and an awareness of unseen forces increases; however, a new spiritually oriented world will not come about without the development of natural gender relationships called patriarchy. Enlightened or not, people need the patriarchal structure in which to function. Chanting OM, Amen or praying the Rosary will not sidestep that need.

I'M BUSY

The statement "I'm busy" permeates our society. Radio and television interviews will invariably elicit the "I'm busy" statement, as will newspaper and magazine commentary. Support groups and self help businesses strive to assist people in coping with busy. "I'm busy" accounts for a lack of sleep, nervous tension, depression, and psychiatric care.

In the midst of all of this "I'm busy" living I decided to share with you the knowledge that I am not busy. I have never been busy, nor will I ever be busy. I am a man and I control my environment. Busy is a condition that I deal with; I do not become it.

"I'm busy" comes from the kitchen. Women are responsive, accommodating, and adaptable; they become their environment. Mom is busy don't go in the kitchen, mom is irritable don't talk to her, mom is late don't bother her. Mom becomes her environment. She becomes the clothes she wears. She also becomes caring, joyful and passionate, which we will come back to later.

Now that women have permeated the work force "I'm busy" has permeated its language. Also, since most children are raised in an environment with a minimal masculine presence they pick up the language and mannerisms of the kitchen. They make excuses for their performance. They were busy. Even more, they now return calls and take action when they are ready rather than when it is appropriate. They act like a woman being courted rather than as a man fulfilling his obligations.

All my life I have returned calls within 24 hours at the latest, and I still do. A few months ago I severed my relationship with a 27-year-old man who was taking care of my computer because he was always "busy." It's not that he wouldn't come and do the work, but he always responded late, didn't answer his messages, and then when I would see him he would indicate "something came up." That phrase "something came up" has been in the female vocabulary as long as I can remember. They are always responding to what comes up—a natural activity for them. What is not natural is for a man to become his environment.

The feminine principle is responsive, that's why we check the weather every day—to see what kind of mood mother earth is in. We don't have to check sunrise everyday—that has been calculated for the year. The Sun is constant; mother earth is inconstant. If the Sun were to become inconstant the solar system would break up. If men become inconstant society breaks up. Have you noticed?

It is one thing to understand that patriarchy is the natural way of life and another to live it. It is one thing to say, I AM A REAL MAN, and to live it. Being a man starts on the inside. You are a center of consciousness around which your world revolves. When you change the inside the outside will respond accordingly. When I lived in Harlem a song I heard sung in the churches was "Working on the inside, changing on the outside." Those words contain a great spiritual truth.

You can't get by with just saying you're a man—you have to act like one. You have to be dependable, constant, reliable, considerate, and respectful. When you do that then you will find women will be caring, joyful and passionate with you. Mother's milk flows freely when she is safe and secure. A woman can only feel secure with you if you conduct yourself in a manner to make her feel that way—firm, yet caring.

You don't have to go to a foreign country to get a good woman; all that you have to do is act like a real man here. I will grant you that most American women are spoiled, but if you want to find out who spoiled them look in the mirror. When you change the image in the mirror you will start unspoiling the women.

The reason I have prepared male Enlightenment and Empowerment seminars is to wash away the feminine habits that our men have acquired and replace it with their natural masculine attributes.

In your relationship with men your word has to be your bond. Women make excuses; men give reasons and there are few reasons that hold up to the WILL to accomplish. Today's society subscribes to the philosophy that if a fly shits on a railroad track, the train stops. That philosophy has no place in Men's Action or among real men.

In my solicitation of members, I look for quality, not quantity. I look to build a core group of real men who will train the quantity of people who will join next year. Those men who conduct themselves as real men will want to associate with their peers. Men's Action to Rebuild Society is a place for this association.

JOBS ARE DESTROYING US

The unnatural repetitive activity that jobs call upon people to perform has created so many human problems that a whole field called occupational therapy has sprung up to deal with them. Every bodily part from the top of the head down to the toes has been subject to severe strain. Doing one application for an extended period distorts the body and produces trauma. Sometimes the trauma can only be relieved by surgery; sometimes it can't be relieved.

Jobs have a negative affect on the mind as well. Most jobs require people to perform a task, which deprives them of the satisfaction of seeing a completed product resulting from their labor. As one autoworker said many years ago, "I'm well paid, but there's got to be more to life than hooking up the chassis to the beginning of the production line." Yes, and there is more to life than putting the headlights in the car, than being a key punch operator, than sewing tote bags all day, than doing telephone sales, than being a checker at the supermarket, than loading and unloading trucks, or adding numbers all day long. Reducing the productive process to its smallest component makes for efficiency; it also makes for the dehumanization of people.

Work on the other had enables people to use their many talents for the benefit of humankind. Original work activity consisted of developing the environment and providing for the needs of people. Men would clear fields, build shelters, plant crops, and hunt for food. They used all their muscles and all their brainpower in the course of their normal activities. Women, while their work was never done, had the healthiest work of human existence. Not only did they use various muscles as they cleaned, cooked, sewed, dressed and suckled their young, they had the psychological satisfaction of seeing the results of their efforts—a clean home, a decorated home, healthy children, and an appreciative husband. They had the spiritual satisfaction of function in an environment that permitted them to give out their love on a continual basis. Constantly emitting love not only does not tire a person, it energizes them. Women of old did not have mental illness; they had very few physical problems and they needed no vacations; why would a woman want to take a vacation from happiness?

Men took from the oceans, rivers, forests, and land what was necessary to sustain the family. They did not attempt to shoot every pheasant in the forest, or catch every fish in the river. They took what they needed. Men worked from Sun to Sun, but they were relaxed and contented.

When men and then women gave up work for jobs their problems began. Instead of working for the needs of the family, tribe, and race, they began working to satisfy individual greed. They peddled their asses for a buck and created a whore society. The whole basis for feminism in wanting the rights of men is that so they could have equal opportunity to make money, Money, MONEY.

This whore society has lost its God consciousness. It does not understand the biblical illustration of the birds that neither sow nor reap but are taken care of. The whore society wants more than that; it wants the rights and opportunity to fulfill its insatiable need for self-indulgence and self-gratification. It does not work for the benefit of humankind. It works for the benefit of self. Consequently all who think like that suffer.

People hate jobs, even though they beg for them. They want the money that the job pays, but they hate the work that they had to do to get it. Tamerlane, the Tatar conqueror said in his sura, "The reward of slaves is the destruction of the fruits of their enforced labor and the impoverishment and humiliation of their masters." I was reminded of this sura when the airline unions had their last strike against Eastern Airlines. A woman striker said, "This strike will probably cost me my house, but we'll show them." It did cost her the house and she did show them. They put Eastern Airlines out of business. The mentality of people has not changed; they want to destroy the fruits of their enforced labor. Jobs are bad for the health–very bad.

In writing this column I thought of an earlier comment made to one of my blogs in which the person said, "Just look at what we (Western society) have done in the last 400 years." I ask that person to reflect on just what we have done. We have built edifices to Mammon throughout the world and have dehumanized the race. Are these not the comments that abound in our blogs and in response to them? We are unhappy, angry, lonely, and dispirited–hardly the makings of a healthy person or society.

Our materialistic and self-centered psyche has created jobs, feminism, and an all-powerful state. Quit your job and do some work. It will improve your physical, mental, and spiritual health. It might even get you to support change, something that will be discussed in my next blog.

LETS STOP TRYING TO FIX THE UNFIXABLE

Would you buy a car that never worked? If the salesman and mechanic explained that it is the best-designed car on the market, it's just that a few kinks have to be ironed out, would you buy it? Would you go to a surgeon who had never performed a successful surgery? Would you go to a doctor who had never treated your type of illness? Then why tote a form of government that has never worked?

The defenders of our republic, Constitution, and democracy offer an infinite number of reasons why it doesn't work. The following were two in response to my last article: "The emasculation of the American male was done through feminist lies, news media lies and incompetence, and judicial activism. A lot of that was both unconstitutional and undemocratic." "The lawmakers and judges in particular have failed the Constitution. Lawmakers pass laws contrary to the constitution. Judges interpret the constitution any way they please, contrary to the intent of the constitution." If lawmakers and judges can't make the Constitution work, who can?

Our conditioning through the educational systems causes us to "believe" that democracy and a representative form of government is the best system, but it has never worked anywhere. The ancient Greeks tried it for a short time and got rid of it. Plato considered it to be the worst form of government before complete tyranny set in. Madison, Jefferson, and Adams all regretted what they had created. De Tocqueville wrote about the failings of democracy in 19th century America, and MND offers an unlimited source of those who make excuses for the failure of democracy.

I offer you Patriarchy. It has worked everywhere! In North American, South America, Africa, Polynesia, and Asia. It has worked for millennia. For the Bible totters, are you not familiar with the patriarchs, Abraham, Isaac, Jacob, and Joseph? Their society was patriarchal, and polygamous as well.

The make up of the central government becomes inconsequential when patriarchy is firmly entrenched. Ghengis Khan, Tamerlane, and the Indian Satraps, all came and went without affecting family life. Patriarchy is an insulator of family life. Patriarchy is the rule of men. The only other choice is the rule of law. I opt for the rule of men any time. The rule of men results in the establishment of family life: the source of our character development, the center of nurturing, and focus of our daily activities.

Those men who are sitting in prison because of unjust accusations of rape or sexual harassment are not interested in some excuse as to why democracy hasn't worked. They want out. Those men who have lost custody of their children are not interested in tinkering with something that caused their loss. They want their children back. Those men who have seen the destruction of all that they lived and worked for through "no fault divorce" are not interested in a dissertation of representative government. Those men who are on probation, who are facing prison, whose lives are in shambles want the hope of something better than that which caused their problems.

We live under socialism that is rapidly devolving into tyranny. I am looking for men—real men to stand up and make change. To be counted. To say that they have had enough, that they have nothing to lose, and that they have everything to gain to make change. I am looking for men who are tired of being classified as wage earners, industrial donkeys, and jackasses on the treadmill of production. I am looking for men who want to reassert their manhood. I am looking for men who are ready to stand up and shout I AM A MAN, I AM A REAL MAN.

Enough with the naysayers and excuse makers, I am looking for men of action. Let me have your comments or contact me at www.mensaction.net. Now is the time to assert yourself.

MODERN FIEFDOMS

The absence of extended family and tribal structure requires the establishment of alternate forms of administration and service to the community. These alternate forms of administration have become known as governments, for they govern the leaderless. Western society has a penchant for developing different forms of government because none of them last or function properly. During the Middle Ages Western man developed the fiefdom as a form of governmental structure. Lords, or liege lords appointed fiefs and gave them a tract of land called a fiefdom. The fiefs, more commonly known as vassals, gave allegiance to the liege lord and served as knights for them in time of war. Serfs worked the land of the fiefdom and obeyed the dictates of the fief or vassal who had jurisdiction over it.

Unlike the bottom up structure of tribal societies, fiefdoms operate from the top down. The rulers of the fiefdom have little contact with whom they rule other than making sure that the serfs fulfill their assigned duties. The fiefs only concern themselves with the well being of the serfs as it pertains to the serf's ability to perform his duties. More often than not, fiefs do not come from among the serfs and have little familiarity with their needs. Democracy and representative forms of government have now become the official types of Western governing bodies, however, fiefdoms reign throughout all their administrative activities.

All government offices and establishments operate as fiefdoms. The officials do not come from the ranks of those to whom they administer, but receive appointments from a bureaucratic office frequently in another geographic local. The officials ((fiefs) get rewarded by their superiors (liege lords) according to how well they do their bidding.

The people to whom they administer (serfs) have no say in anything that the officials decide. The people have no other office to go to; competition does not exist in fiefdoms.

Government regulated private industry such as utilities, cable, transit, and telephone companies, all operate as fiefdoms; they have no competition. They get their appointment from the government (the liege lords), are operated by officials (vassals) who do not come from the ranks who use their services (the serfs).

The proliferation of non-profit organizations that permeate our society serve as classical examples of fiefdoms. The government and charitable institutions that charter and support non-profits act as the liege lords, the officials of the non-profits act as vassals, and they treat the people who they administer to as serfs.

The bureaucratic nature of private industry frequently devolves into fiefdoms as the primary concern of middle management revolves around protecting their turf while fulfilling the wishes of top management. Middle managers (fiefs or vassals) administer to the workers (serfs) and do the bidding of their lords (top management).

Most religions operate as fiefdoms. The ruling oligarchy sends vassals to the various houses of worship to attend to the need of the serfs. In some religions the congregation has the right to disapprove of the vassal sent to them, and in some cases can have him removed; however, only on the rarest occasions can the congregation suggest that one of their own act as a vassal. The primary activity of all religions once centered around propagating the faith, but that activity has shifted to maintaining control, which is not done by catering to the congregation but by catering to the government–the supreme lord–that bestows non-profit status among the official religions.

Fiefdoms stifle innovation, impede creativity, and emasculate its members. They do little if anything to develop the character and ability of those they ostensibly serve.

Fiefdoms are control vehicles. All forms of government developed by Western man are control vehicles. There is no form of government that man has devised that works for the development of people. Western man does not think of the development of people; he thinks of the satisfaction and glorification of self. He does not have a tribal mentality or a race consciousness; he has a self-consciousness. He is narcissistic. He measures himself by what he owns and controls.

God or nature established the family, tribe, and patriarchal structure that make them possible. Its members concern themselves with the propagation of the species and its spiritual development. Tribal chiefs care about their members. Fiefs care about themselves.

Western society has developed a national angst. Its people are lonely and frightened; they see society imploding upon itself. Who can they turn to for solace? Who cares about them? Who even knows their name?

All forms of government devolve into fiefdoms, the only difference between them being one of degree. The fiefdom results from the lessening of the masculine influence and the incorporation of the feminine lack of government and structure. Men build groups. Women have cliques. Women's clique activity starts in pre-kindergarten and continues through all levels of school, into industry, their private lives, and until they die. Governments that incorporate the clique syndrome are known as aristocracies; their objective is to keep others out.

Aristocracies have their place; they level people up. People want acceptance and in order for it to take place in an aristocracy people must excel in the activities that it deems commendable. The clique of housewives leveled up the cleanliness of homes and care of children, for dirty house-keepers and those with unruly children were not admitted into the clique. However, unless a masculine influence tempers the aristocracy it will self-destruct; one reason that all purely female organizations ultimately collapse.

In a subsequent blog I will describe the credentialization of the educational system and other areas of society as aristocratic practices proliferate.

MUSINGS ON THE RELIGION OF SCIENCE

Science is man's somewhat organized record of his ignorance of himself and the universe in which he lives. It is a direct outgrowth of the European, feminine, materialistic culture, which does not understand anything that it cannot see. Aristotle, one of the earlier practitioners of what has become known as the scientific approach, gathered, analyzed, classified, and recorded. He amassed information on a multitude of natural manifestations such as seashells, rocks, fish, and various flora and fauna. He used deductive reasoning in conjunction with his analysis to get to the truth or essence of a particular matter; however, he never got to the truth. The European philosophers that came after Aristotle all searched for the truth but they never found it.

To compensate for their lack of knowledge, the European intelligentsia came up with "theories", which were nothing more than guesses as to what might be the truth. These theories, when repeated often enough become accepted as fact and then become revered. This reverence for what is ignorant is akin to the practice that existed in China about a century ago when the feet of Chinese girls were bound so that they could not grow to normal size, in order to emulate the feet of the emperor's daughter, which were deformed. An abnormality was held up to as something to strive for, much as the ignorance of science is held up as something to strive for. When science comes to the realization that it has an erroneous theory, it comes up with a new theory that it heralds as a scientific breakthrough.

One such area of changing theories was in regards to man's understanding of matter. When in high school, part of my physics instruction indicated that matter could neither be created nor destroyed. As proof of this theory it was determined that if something was burned, the total residue of the fire plus the total of all the gases emanating from the fire would equal the sum total of matter of the object that was burned. This was a typically European and feminine theory; a god was made of matter, an entity that could neither be created nor destroyed.

I compare this scientific theory with what was written by a mystic in 1904 in which he stated that science considers matter to be neither creatable nor destructible, but that within 50 years it will understand that matter is but a form of energy. He went on to say that those who would be born in the early part of the century would chuckle when they read this, because they will have witnessed the change in scientific theories. I was one of those people who chuckled for at the very time that I was in my physics class learning the theory of the indestructibility of matter scientists were working to develop the atom bomb, a manifestation of the converting of matter into energy. After the atom bomb was developed and used to blow up two cities, the physics courses were changed to include a more "advanced" theory about matter and energy, which stated that matter was a form of energy, and energy could neither be created nor destroyed. The mystical text went on to indicate that eventually science would learn that energy is but a form of mind.

The comparison of the mystical and scientific understanding of matter and energy is equivalent to the comparison of knowledge to ignorance, of the unseen to the seen, of the masculine principle to the feminine principle, and most importantly of the understanding of the oneness of all existence with the concept of independence and separateness. A whole universe of knowledge exists outside of the physical one that science classifies and records, and there is nothing wrong with using the scientific approach to learn what makes the physical world tick, after all it did enable European man to learn (belatedly) that the earth rotates around the sun; however, when it exalts its ignorance and puts down the learned, science becomes a dangerous weapon in the hands of the few. Its ignorance becomes arrogance and its belief system becomes a religion enforced by an aristocratic and credentialized hierarchy.

The ignorance of the scientific approach can be found by logging on to Google and looking up a subject of your choice. There you will find scientific viewpoints that disagree with each other, frequently on the same page. Two subjects I looked into were the sperm count of the American male, and global warming. Under the first category I found papers written indicating that the sperm count has dropped, some said by 25%, some by 35% and some by 50%. On the same page I found a scientific report indicating that there is no evidence that the sperm count had dropped at all in the past 50 years. In regards to global warming, I found reports indicating the severity of the situation caused by global warming and the catastrophic events that could occur because of it. I also found reports indicating that global warming was a scientific myth. If the results of scientific studies can be so contradictory, the scientific approach should be suspect; instead, it has become a highly revered religion.

Do we not see daily in the papers or hear on the radio or television what science says about this or that? To even use the term science as a noun gives it an identity, a sense of being. Like all religions it has its hierarchy. Scientists write papers, books, and articles, which gives them ranking in the pecking order of their specific field. The universities they went to and the numbers of degrees they possess are used to measure their stature in the community. They form sub-sects of their religion according to their field of study, have membership, and conduct symposiums in which new theories are aired or old theories reinforced. We become so awed by the priesthood of science that even the government invites them to address various congressional committees.

There is nothing wrong with showing respect for science, the thing that I caution you about, is not to be awed by what scientists say. They are ignorant people groping for the truth, and all too often they enjoy basking in their own ignorance. Knowledge does not come from scientists, or from universities, or from the government. The only source of knowledge is the Godhead. It is transmitted through the masculine principle and then materialized and personalized by the feminine principle into worldly information. We live in the information age, not the knowledge age, and certainly not in an age of truth. The tendency of the Western culture to see everything in material terms and then make a religion of their belief system tends to draw all of us into its web of ignorance.

NON-PRODUCTIVE CAREERS

A review of career listings found in various locations on the internet and in print media reveals an increasing number of professions and occupations that have no purpose other than to compensate for the un-natural lifestyle that we lead. Careers such as urban planner, financial analyst, health services, human services, recreation worker, economist, government and public administration, agricultural and food scientist, and public safety, corrections and security, are all creations of Western society. Natural societies had no need of these careers.

Why do we need agricultural and food scientists? God made food and designed a certain way for it to grow and be harvested. God's work cannot be improved upon. The abandoning of natural farming for modern productive efficiency destroys the natural balance of nature and requires the employment of scientists to figure out ways to produce food from an abused environment.

What do we need economists for? Productivity and the amassing of things material has become a fetish of modern society. God has given us freely all that we need. Paul said, "There is no single thing that you have not received." What function do economists fulfill? They have generated a multitude of theories because they know nothing. Their predictions on economic growth are no better than guesswork.

Public safety, corrections, and security are only needed in a society lacking in ethics. Ethics come from the masculine principle and become imbued within society by family training. Jails did not exist in the Americas, Polynesia, or Africa before the arrival of Western man. Western nations all have huge penal and security systems to protect themselves from each other because they have destroyed the fundamental structure of humankind—the family.

Vocational activities such as occupational therapy, family counseling, childcare, elderly care, health care, investment counseling, legal services, social work, and drug and substance abuse treatments are all unique to Western society.

The un-natural activity of working at a job created the need for occupational therapy. Homemakers of old in all cultures didn't need occupational therapy because they used all their bodily parts in the course of a days work. They lifted, carried, stirred, mended, scrubbed, ironed, peeled, laundered, and did a variety of other chores that exercised their entire bodies. Men, would plow, make fences, chop, hone, hunt, fish, and do a variety of other activities that would utilize all their muscles.

Jobs have people sit or stand in one position all day; they go through the same motions all day putting undo strain on their bodies. Their minds also become numb. They need physical therapy, mental therapy, and various supports for their bodies. Jobs are unnatural. People were not created to have jobs. They were designed to do whatever was necessary to propagate and preserve the species. In natural societies, if a child was sick and a mother had to go through the night without sleep taking care of it, that was her responsibility. If a man went on the hunt for food in dangerous terrain at the risk of losing life or limb, that was his responsibility. Men and women did what was necessary to preserve the family and tribe; they did not have jobs. Work is good; jobs are bad.

Jobs have become necessary in order to obtain money so that the essentials of life can be obtained, essentials that man once provided for himself before all the resources around him became "owned." Jobs are demeaning, dangerous, and life threatening. Jobs are economic slavery. Yet the slaves today are begging the government for more jobs, so indoctrinated have they become to their condition of servitude.

Family counseling represents another vocation brought about by the aberration of Western society. Young college graduates certified as social workers call upon women with children and tell them how to be good mothers and maybe wives. These functions were normally learned from grandmothers and great-grandmothers, aunts, sisters, and cousins. Learning how to become a good wife and mother is not obtained from a book; it comes from observing how others do it. Family provides the environment for that education to take place. With the destruction of the family the center of learning for parenthood disappeared and family counseling was created.

Space does not permit going through each and every career and occupation developed as a direct result of the unnatural practices of Western society. However, the examples given will hopefully serve as a catalyst to explore all that we do that has no direct contribution to our well-being. Living in such a manner leads to stress, anxiety, angst, and unhappiness. It leads to spiritual as well as material poverty.

Do you want to continue living in this manner knowing that conditions will continue to worsen; or do you want to learn more about these issues and perhaps become involved with others who have similar concerns? You can receive these essays on a regular basis at no charge by clicking the subscribe button under my picture on www.mensaction.net. You can also subscribe to my videos at no cost through youtube.com. Should you become further interested and want to help share the word you can invest in my book and information booklets through paypal.com on the Men's Action website.

Getting back to the conclusion of this essay, an increasing amount of our labor goes to support the payment of the consequences of the un-natural lifestyle that we have developed. You might want to refer to my extensive essay entitled, *The Fixit Society,* or, to learn more of the conditioning we receive for this un-natural lifestyle you can review *Cubists.* Each year we have to work harder to maintain the lifestyle we had the previous year because an increasing amount of what we earn goes to support non-productive activities and a growing government.

Government continues to expand its control into every aspect of our lives through overt and covert activities. The printing of money to stimulate the economy will result in inflation that will erode the value of social security benefits and retirement income. Ultimately we will all become wards of the state. It will decide on our health care, housing, education, recreation, and perhaps the length of our lives as well. In the process it will create more non-productive jobs necessary to for the administration of our control. You can do nothing and await the furtherance of these conditions, which are already upon us, or you can decide to get together with others and start to develop a way of life outside of the system. Men's Action offers you a vehicle to do the latter.

OUR NEGLECTED AND FORSAKEN YOUTH

While riding on the subway this week I noticed a woman sitting next to me reading a report concerning the behavior of preschoolers. I mentioned that the main issue with preschoolers was that they didn't have mothers at home to provide the love and nurturing that children that age require. She challenged me and in a militant manner but the subway had reached my stop and we couldn't continue the discussion.

That event reminded me of the comments made by a Kenyan woman who I know regarding Western thought. She said wherever Western man goes he disturbs what is natural and then creates a "fix it" mentality to deal with what he has disrupted.

Her viewpoint becomes readily apparent upon addressing the condition of our youth today. At least one million of our youth belong to street gangs. Girls at the ages of 10 and 11 have sex clubs in grade school, even in parochial schools. Pregnancies now occur to 40% of teenagers. A government statistic indicates that America has the most violent boys in the world. These are but a few of the many illustrations of decadent conditions that prevail among our youth.

The cause of these conditions has been the destruction of the natural family environment necessary for the development of children. Ninety percent of gang members come from broken homes, as do 90% of the men in prison. Twenty four million children go to bed at night without a father in the home. Only 12% of our 18 year-olds live with both their natural parents. For all practical considerations the family has been destroyed in America and the effects can be readily seen among our confused, lonely, and misdirected youth.

Getting back to the woman on the subway with her report on the behavior of preschoolers, the raising of children by the state is not something of recent development. In his *Republic,* Plato calls for the turning over of all babies to the State with the resultant abandonment of the family structure.

Government control of all aspects of the life of its citizens is inherent in Western thought and is contrary to natural order. Whatever is contrary to the natural order deteriorates; our children were raised in an un-natural environment and their condition has deteriorated.

Our children require love and nurturing when they are young and guidance and discipline when they are older. The natural way for them to receive this is through their mothers and fathers. We have denied our children the warmth and security of the home. They have been forsaken. Most of you who are reading this have been forsaken. The deliberate breakdown of the family began more than two generations back and it has affected most of society.

Girls no longer understand why they were born females and boys do not understand why they were born males; neither knows their purpose in life. They no longer receive instruction in manhood and womanhood. They no longer dress and act in a manner to attract the opposite sex for the purpose of mating and producing offspring. All standards have deteriorated. Our youth just hangs out together with personal gratification being the only objective or even reward, which is temporary at best. I recall reading an article in which a woman college graduate said, "What's a date?" She had not been on one date during her time in college—a very sad situation. People have no sense of purpose; that makes for a dull and impotent existence.

The government no longer wants you to behave like men and women but as persons, as automaton production workers and directed consumers. While you do these two functions the government takes over every aspect of your lives. The government might project the image that it cares about you, but only people can love you. Usually the people who love you best are your parents or members of your extended family, but I realize that is not always the case.

When we address the root cause of our forsaken children their condition will improve. The root cause is the absence of the patriarchal structure, which in turn resulted in the destruction of the family. Patriarchy is family. Men make patriarchy. No men—no patriarchy. No patriarchy—no family.

I ask you all to consider this statement. Either man rules or the government rules, there are no other options. Our forsaken youth are a result of government rule. Don't you think its time for men to reclaim their manhood and reestablish the family as the focal point of society?

If you do, then review our mission statement and become involved in the promotion and dissemination of our message.

To the women who are reading this article, while I call upon the men to make change, I call upon you to support them and nurture them. They are risking their lives to make a better world for you and your children.

RAISING CAIN

New York's Channel 13 ran a documentary entitled Raising Cain, which stated among other things the discomforting statistic that America has the most violent boys in the world. Cain was Adam's first son and slew his brother Abel, hence the term *raising Cain*, which is used to refer to creating violence. This documentary addressed the various stages in the life of boys from birth to adulthood, and surprise surprise, they concluded that boys are different than girls but they didn't seem to know why. The ignorance of Western society concerning gender knows no limits.

The research that went into this documentary and the cost of its production required the expenditure of large sums of money, and for what purpose? The documentary reached the conclusion that boys and girls benefit when educated separately, a fact that the non-Western world has known for millennia. Ignorance of gender difference has not only created the most violent boys in the world, it has also created the highest incarceration rate in the world. America has more men in prison than any other nation. It has the highest rates of unwed motherhood, and debilitating depression among women. These statistics give evidence of the malfunctioning of Western society, which I believe is brought about by the lack of understanding of gender and is the reason that I continually harp on this issue to my readers.

All of life depends on the family structure and is the reason that the Torah, Bible, and Koran address family frequently. More than 90% of men in prison and boys in street gangs come from broken homes. Raising Cain pointed out that boys who are raised in homes without fathers tend to become much more violent that boys who come from in tact families; yet, little is done by our government to support family and much is done to destroy it. Those few who promote family tend to teach fatherhood as though it were a job description like carpenter or dentist; not realizing that good fatherhood comes from developed manhood. The structure that best develops manhood–and womanhood also–is patriarchy; a societal structure inherent to the universe.

The animosity between men and women in the Western culture is unnatural and does not occur in non-Westernized societies that focus on the group, it preservations, growth, and the interdependence that makes that activity possible. The focus of the Western culture is on the individual and self-aggrandizement; consequently, the relationship of the sexes takes on a different connotation than it does in non-Western societies.

The propagative efforts of these non-Western societies are part of the activity of the universe, which pulsates with energy and movement as everything in it from the tiniest atom to the largest nebulae continually expands and reproduces itself. Life exists everywhere and in everything; nebulae spin off new suns, suns spin off new planets, and planets develop new life. Just as space teems with activity, so do the forests, which vibrate with a symphony of distinctive sounds created by insects, birds, and mammals as they go about their reproductive activities. The wind blowing through the trees and rustling through fallen leaves, the creeks and rivulets rushing over rocks and stones, the falling of decayed branches, and the thunder of impending storms all add to the sounds of the forest; sounds that result from the creation and care of life.

The sounds of mating rituals, the activity of building shelters–whether nests in the trees or holes in the ground–the care and feeding of offspring, and warnings of impending danger, fill the air. Some insects burrow in the ground, some slither and crawl along the surface of things, and some fly as they all go about their activity of bringing life into this world and sustaining it. The forest provides protection for brush animals such as squirrels, skunks, rabbits, raccoons, hedgehogs, and foxes. It provides sustenance for deer, bear, moose, and elk; and living accommodations for woodpeckers, hawks, blue jays, robins, sparrows, owls, and the myriads of other species of birds. The forest contains many civilizations, all of them interdependent, all of them working feverishly–sometimes to the point of exhaustion–to propagate and preserve the species.

Deserts–barren and lifeless as they might appear–also contain various forms of life that strive to survive, propagate, and preserve. The desert cacti, the mountain freshet, and the occasional well hold the precious water that draw other forms of life to it. Insects, lizards, snakes, birds, and animals suited to the arid conditions work at reproduction and survival, and in the process produce their own forms of beauty.

Oceans teem with life and activity from its bottom where the monera–the simplest form of life–abounds, to the surface, where the world's largest mammals swim about. The oceans contain plant and animal life of every shape, size, and color forming societal groupings and civilizations all of which work at reproducing and preserving themselves.

The world vibrates to its reproductive rhythms, whether from the ocean, the land, or the air, and these activities produce works of beauty everywhere. When we admire the beautiful flowers of the fields, birds of the air, fish of the sea, and animals of the land, we are looking at the manifestations of the earthly activity of reproduction. Humans add to this reproductive beauty whether in the form of a young, beautiful, and graceful woman or young, strong, and handsome man. There is beauty in the newborn child, and a sense of happiness in observing it, for it makes us feel a part of the great universal life and purpose; we inherently feel the joy of fulfilling the great command, "Be fruitful and multiply."

The principle of gender makes possible the great reproductive activity of the universe, and the next series of blogs will describe this principle in depth and detail. Living in accordance with the universal principle of gender results in harmony; whereas, living a life ignorant of this great principle results in the degeneration and death of society.

SEXUAL EDUCATION IN OUR SCHOOLS

Families have lost control over the raising of their children in the same manner that craftsmen have lost control over the manufacture of their products.

Craftsmen used to produce and take responsibility for the performance of the products that they made such as a wagon; they even signed their name to the their products. The development of the assembly line, which reduces all work operations to simple tasks and then standardizes these tasks into a productive routine, eliminated the craftsman. No one worker has the responsibility for the completed product nor does he normally see it. Western society considers the replacing of craftsmen with production workers who do single tasks to be part of economic growth.

Taking away responsibility for ones accomplishment now extends to the home as well, especially as it pertains to the education of children. Parents used to educate their children in all that they needed to know to help the family survive economically and morally. Parents were the craftsmen of the home and taught their children how to cook, clean, milk cows, hunt, chop wood, plow the fields, make clothing, and a myriad of other activities that provided for the family's material needs. Cultural practices provided for the propagation of the race through marriage. A set of ethics and/or religious practices provided for the moral development and conduct of the family and tribal members.

Compulsory public schooling has replaced the influence of the home as well as all the standards that the family lived by. Children no longer receive an education that has anything to do with the preservation of the family. Education now focuses on training children to be obedient production workers and automaton consumers. They learn of a life of self-indulgence and debauchery.

The constant mingling of the opposite sexes in the school system provides the opportunity for extreme licentiousness. On Sunday May 23rd 2004, The New York Post in a front-page story broke the news of sex clubs in which schoolgirls wore sex bracelets that advertised the type of sexual activity that they would perform. For instance, black indicated sexual intercourse, blue meant oral sex, and glow in the dark meant using sex toys. A whole menu exists of the sexual activities the girls offer. The term girls is quite literal, they are 11 years old!

The picture on the front page showed a Catholic school that hosted one of these sex clubs. Girls in public schools, private schools, and religious schools participated in these activities; the democratic nature of immorality does not show preferences. Outside of the one-day splash by the media, I have not heard this subject brought up again. America has become inured to this type of behavior, for it is a natural consequence of educational policies that promote moral relativism—the concept that there is no right and wrong—which in turn leads to conduct based upon what is legal and illegal rather than what is moral and ethical. The extent to which moral relativism has infiltrated the educational system and our societal fabric is explained in a book written by Tammy Bruce entitled, *The Death of Right and Wrong.*

Miss Bruce informs us that the educational system under the guise of sex education really teaches sexuality. It not only teaches about sex, but how to do it. Under the heading of Alternate Lifestyles it not only teaches lesbianism and homosexuality, but how to participate in it. Since there is no right and wrong sex becomes an object of pleasure, as do all things in a society that has no moral basis. Teaching children sexuality at an early age increases their sexual activity and results in teenage pregnancies, increased incidences of venereal disease, mental confusion and depression, and an increase in suicides among girls.

The June 2006 issue of Family News from Dr. James Dobson indicated that New York City's Department of Education recently mandated that all five-year-olds must learn about AIDS and HIV in their kindergarten classrooms. Can you imagine teaching five-year-olds about sexual transmitted diseases? I went to the library to verify the article that Dr. Dobson referred to and he was right. This past September New York City teachers began teaching AIDS and HIV to five-year-olds!

An article written by Charles Millard entitled Replacing Parents appeared in the June 19th 2006 issue of the New York Post and focused on Mayor Bloomberg's program to provide contraceptives and morning after pills to high school students. Parental consent or advisement is not required. This action shows a complete ignorance of the purpose of sex, a disregard for the moral standards of the parents and of society in general, and a motivation for teens to become less inhibited in their social relationships.

The activities that break down the moral fiber of our youth seem endless. There is a bill under consideration in both houses of the New York State Legislature entitled, Dignity For All Students Act. This is a pleasant and positive sounding title, but it masks the intent of the bill, which mandates the teaching of homosexuality to all public school students from kindergarten through 12th grade. This bill attacks the very purpose of our earthly existence.

Its ostensible purpose is to end the discrimination and harassment of students because of race, religion, gender, disability or sexual orientation. However, the Dignity For All Students Act is an attempt to normalize homosexual behavior, and usurp parental moral authority. It is the equivalent of what some educational institutions call the teaching of "alternative lifestyles." It has another dangerous aspect, making remarks that indicate women are different then men would be the equivalent of a racial or ethnic slur; by law we will be required to consider women the same as men, which is at the heart of all the mess in the educating of our children.

If we believe that the purpose of sex is to procreate the race so that we can grow spiritually, then offering homosexuality as an alternate lifestyle is contrary to this purpose, as is the teaching of pre-marital sex. However, the Western culture believes in the freedom of the individual to do as he chooses, without having any obligation to the family, tribe, or race. This viewpoint results in the destruction of family its race preserving values. Legal and illegal supplant right and wrong. As the prophet Isaiah stated, "The wisdom of their wise men shall vanish and the discernment of the discerning shall be lost."

Three million American girls suffer from depression and American boys are the most violent in the world. America is destroying its youth and that activity will ultimately destroy the nation.

We cannot depend upon democrats, republicans, liberals, or conservatives to make change; they have all contributed to the catastrophe engulfing our youth. It is time for Western man to face up to his responsibility of providing the environment and means for women to bring life into this world and nurture it. It is time to provide security for our children and the moral guidance necessary to preserve the species. It is time to stop worshiping the work of our own hands and realize that we have been given all that we need. It is time to realize that just maybe we can learn something from someone else and collectively bring about the needed change.

STAMPING OUT FEMINISM

> *There are people in Europe who, confounding together the different characteristics of the sexes, would make man and woman into beings not only equal but alike, They would give to both the same functions, impose on both the same duties, and grant to both the same rights; they would mix them in all things–their occupations, their pleasures, their business. It may readily be conceived that by thus attempting to make one sex equal to the other, both are degraded, and from so preposterous a medley of the works of nature nothing could ever result but weak men and disorderly women.*
>
> Alexis de Tocqueville

An attempt to stamp out feminism will have less of a chance of success than an attempt to stamp out cold; neither of them exists. The absence of heat produces cold just as the absence of light results in darkness and the absence of sound results in quiet; however, there are no entities such as cold, darkness, and quiet; they are the effects of a lack of heat, light, and sound. There can only be one assertive influence causing any condition, if there were two, constant conflict would ensue. If darkness were an entity it would constantly fight light. The same conflict would occur if quiet or cold were entities. The attempt by government to legislate the receptive feminine entity into an assertive influence has created such a conflict. The effect of this legislation has produced an effect called feminism.

Feminism is an effect such as malnutrition, diabetes, and cirrhosis of the liver; they are all symptoms, not entities unto themselves. Feminism has been likened to evil, which is also an effect, it is the absence of good. No amount of legislation can eliminate the evil that ensues from the absence of good. The United States has more lawyers per capita than any nation in the world and more laws; it also has more men in prison. Our nation has a high incarceration rate and high rates of various types of crimes. It has a high level of teenage pregnancy, unwed motherhood, divorce and adultery. Drunkenness, debauchery, and crime continue to grow in all areas. We have reduced our ability to create good. The greatest societal good comes from the family and there can be no family without patriarchy. The legislating away of the power of men has destroyed the cohesive and assertive power necessary to build family. The government has taken away all authority of men and reduced their function to that of a work unit. Consequently the family has collapsed and all morality along with it. The United States has the most violent boys in the world; putting more of them in jail has not decreased the violence, it has increased it.

The laws that the government have created are an effect of Western man not having an understanding of the universal patriarchal order nor an ability to implement it on the rare occasions that he does understand it. His thinking is feminine oriented and like a woman he too considers himself to be a victim of all things. Poor Western man, his life has been torn asunder by those bad feminists. To address his grievances he goes to the very government that created the conditions that he complains about, an action similar to a slave calling upon the master to improve his working conditions; it doesn't happen.

By removing the power of men the government has done the equivalent of removing the sun from the solar system. Without a sun there will not be any energy for nurturing to take place, the planets will fall out of orbit, chaos will ensue, and the solar system will collapse. Likewise our society has lost its virility, nurturing has decreased, women have lost all sense of propriety, and our society is collapsing. No amount of government intervention can change this condition. Only virile men who will rise up and assume their familial responsibilities can create the conditions where nurturing can take place and good will develop.

THE BOY CRISES

The January 30th issue of Newsweek contained a cover story entitled, The Boy Crises, which was written by a women, had as its first example an issue of an unwed mother, contained a photo of boys serving detention with a woman seemingly in charge, and at the end treats the reader to a full page article from a feminist scholar (an oxymoron). Understandably the editors did not see anything strange about this presentation, as the following flashbacks of Newsweek covers will illustrate:

Newsweek's June 2, 2003 cover story entitled America's Best High Schools had on its cover the picture of a girl student. The story inside began with a picture of another girl; the school selected as its example was an all girls' high school in the Bronx, and many of its students earn extra credit by doing community service at a home for unwed mothers. Lest anyone think that Newsweek made a mistake, on May 16, 2005 it again ran a cover story on America's Best High

Schools and again the student on the cover was a girl. The photographs accompanying the story contained a few shots of girls only, but no shots of boys only; apparently boys were being accommodated into the world of girls.

No matter what the national or even international thrust of its cover stories, the photograph on the cover will be of a woman whenever possible. The Newsweek cover for May 9, 2005 featured China's Century lettered across it with a picture of a woman representing the new China. I guess there are not only no boys in high school, there must also be no men in China. The May 28, 2001 cover contained the caption The New Single Mom—Why the traditional family is fading fast, accompanied by a picture of a mother and her son. Of course the traditional family is fading fast, there can be no family without men. Further evidence of the breakdown of the family can be seen on the cover of the July 12, 2004 issue of Newsweek, which featured The New Infidelity--More wives are cheating too. Again the picture on the cover is of a woman. Newsweek also made a special effort to denigrate the American born Black male with its March 3, 2003 issue, which states on the cover, From Schools to Jobs, BLACK WOMEN Are Rising Much Faster Than Black Men.

The Newsweek cover story The Boy Crises, while feigning interest in the education of boys really served to further denigrate them, an activity which as we have seen, has been a part of their editorial policy for years and is part of the general media efforts to emasculate the male. US News & World Report has done equally well in propagating the distortion of gender differences.

US News also had its cover story about American high schools entitled Outstanding American High Schools in the January 18, 1999 issue in which three students grace the cover; one black, one white, and one Asian; however, none of them are boys. Like Newsweek, I guess US News believes high schools are essentially female institutions. They did not keep us guessing for long, for in their February 8, 1999 issue the caption on the cover is Where the Boys Aren't, meaning college. The cover has an illustration of eight college students in their graduation robes and only one of them is male. The March 22, 2004 issue's cover contained the caption Unequal Education---Why so many kids are still being cheated and contained a photo of at least six grade school children-all of them girls. US News took us from high school through college to grade school with an almost total absence of the male sex on its covers. The trend continued with an advertisement for their August 29, 2005 issue stating, This is the issue to be in! America's Best Colleges. A picture of a woman college graduate accompanied the ad.

The emphasis on the female and the relative absence of the male on the covers of these major news magazines is not coincidence or oversight; it is a concerted effort on the part of the media to eliminate the masculine influence, breakdown the family, and establish a completely socialistic government. US News provided proof of this viewpoint in an article written by Jodie T. Allen entitled, "Are men obsolete?" that appeared in the June 23, 2003 issue. Some of her statements are; men are a perennial and perhaps deepening problem; boys dominate in such dubious categories as remedial education, stimulant-drug prescriptions, and suicide; and there remains, to be sure, one large sector in which men retain unquestioned domination: crime. She asks, "What shall we do with all the men?

Jodie T. Allen was not some disgruntled reader who wrote a letter to the editor, but a paid member of the staff of US News & World Report. She is on the Zuckerman payroll. Thee Mortimer B. Zuckerman, Editor-in-Chief of US News, owner of The New York Daily News, real estate magnate, and billionaire. A man with these credentials cannot claim ignorance; he knows the effects of lambasting the male in his national news magazine. All the cover stories and this article written by Jodie T. Allen are part of a strategy to breakdown the family.

No outside force has a devastating effect on our way of life that even comes close to the destruction caused by the liberal media. Each week the combined efforts of the national news magazines, television, radio, and internet bombard us with messages aimed at breaking down the natural relationship of men to women, which in turn breaks down the family, which in turn develops the need for government intervention into every facet of our lives, which leads to ultimate control of everything by the government. Make no mistake, Mr. Zuckerman and his cohorts are evil. They are the greatest evil facing this nation. They are molding our minds in a manner to suit their purposes. What they refer to, as "freedom of the press" is an unlimited and open-ended opportunity to propagandize us with grossly materialistic values known as secularism.

THE ENEMY

The enemy surrounds us and also dwells in our midst. The enemy affects every aspect of our lives; the water we drink, the food we eat, the air we breathe, the thoughts we think, the money we earn, the money we spend, the products we buy, the services we need, and the services we get. It controls our education, entertainment, and health care. The ubiquitous enemy has the qualities of a god, it seems ever present, knowing, and powerful.

To fight this enemy is futile. It has written the rules of engagement and decided upon the arena in which any battle will be fought. It controls the land that provides our food. It controls the lakes, rivers, and watersheds that provide our water. It controls our fuel. It completely controls our environment. It has tanks, planes, arms and artillery. It can attack us from the air, land, and sea. It controls the media and all communications and can maintain sustained psychological warfare.

What then can we do to protect ourselves from further encroachments of our liberties? What can we do to gain some measure of control over our lives? Can this enemy be brought to heel?

The enemy has an abundance of abundance. That accounts for its strength and also its weakness. The enemy has put its faith into what it sees, but has no grasp of what it cannot see. It's ignorance of nature and the unseen has caused it to sow the seeds of its own destruction. It has polluted its environment, destroyed the family, overspent its finances, forsaken its youth, and neutered its men. It does not have to be defeated—it has created conditions that are leading to its own demise. It's heartening to have an enemy whom we know will defeat itself. However, if we wait for the enemy to self-destruct we will get sucked into the undertow of its sinking.

Our only rational option is to live our lives outside of the value structure of the enemy. We need to live in harmony with our understanding of the unseen and with nature, as individuals and collectively. We need to live our lives that we were created to live. The structure for that society is patriarchy; it enables development of the individual family and its evolvement into the extended family and then the tribe. It requires that people work together to support the common good. It fulfills our ordained responsible to propagate and preserve the species while we grow in God's love. It is a way of life that all religious texts direct us to live.

Before patriarchy can be established, or even a life started outside of the mainstream, men must first become empowered. They must come to realize the power within, the purpose of their existence, and the responsibilities of their lives.

THE EUROPEAN PSYCHE

The European psyche believes in unbridled freedom, self-indulgence, material acquisition, and a self-important nature that results in aristocratic societal structures. These are feminine characteristics, which when in balance with masculine characteristics become attributes, but when left unchecked lead to moral decadence, chaos, and a strong central government controlled by an aristocratic elite. We have such a government now, and for the most part that has been the nature of the governments of the West for the past 2,500 years.

We have moral decadence on a scale that encompasses every facet of society, and every profession including the government itself. Well-paid heads of large corporations embezzle billions; employee theft amounts to more than a billion dollars a week, and consumer theft known as shrink, amounts to one and a half percent of retail costs. Officials in sports, business and the professions accept bribes, as do members of government who legislate and enforce the law.

Marital infidelity and divorce run rampant among the populace and clergy as well. Children spend inordinate amounts of time in nurseries and after-school care, in orphanages or foster care. Elders are put in senior citizen homes. Young people live separately or cohabitate without marriage. The majority of adults are now single. Family—the prime source of values and character building—has been destroyed and right and wrong have been supplanted by legal and illegal.

Religion has taken precedence over spirituality. Church dogma supplants spiritual understanding and religious history (a record of the seen) takes precedence over the unseen meaning of scripture.

The Western insatiable desire to accumulate things governs all its motivations and its relations with others from individuals to nations. People go to school, to get an education, get a job, make money, and be somebody. They do not go to school to learn; instead they go to acquire information that will enable them to amass material things. At the national level government measures societal well-being in terms of gross domestic product—a barometer of how much consumable material the nation produces.

The Western psyche's desire for material accumulation results in the need to "own" everything, and land consists of that which it can own. Wherever Western man has traveled internationally he received friendly greetings from the local inhabitants who contributed to his survival. He received this hospitable treatment form the indigenous peoples of North and South America, Africa, the Middle East and the Far East. In all instances of this hospitality, Western man took ownership of the lands of his hosts and made the people slaves, serfs, and indentured servants.

The Maoris in Hawaii exist at the bottom of the American economic ladder. In South America the indigenous people are referred to as those "Indians" and have little influence in the operation of their governments or economies. In North America those "Indians" live on reservations. The tribal ways of the African have for the most part been destroyed and Western interests control the land. Western man has divided Asia into two economic environments of his design—communism and capitalism—both of which have destroyed the cultural and social structure of their lands in the process of opening them up for economic exploitation. The destruction of cultural and religious practices in the lands he occupies results from the aristocratic aspect of the Western psyche and led to the segregation of non-Westerners all over the world.

In his insatiable desire to own anything and everything, Western man has destroyed and polluted the environment to a degree unheard of in all of recorded history. He has polluted the air he breathes, the water he drinks, the food he eats, and the thoughts he thinks; and considers this pollution the price of economic well-being. He shows an open willingness to sacrifice health for money.

This sacrifice of health has caused illness of epidemic proportions. The extensive use of prescription medications to treat these symptoms has caused economic hardship on the populace and is creating strains on the federal health care budget.

When Western man points with pride to his achievements, he does not realize that they consist merely of the rearrangement of the earth's material done at the price of disrupting natural harmony. The continuous deterioration of all that he created requires continual maintenance.

The gross materialistic thinking of Western man has so blinded him to the natural functioning of the universe that he does not know how to live in harmony with it. His society crumbles before his very eyes, and he knows not why.

Western man does not realize his responsibility regarding the propagation and preservation of the species, an activity that cannot be accomplished outside of the patriarchal structure, which in turn requires a clear understanding of the principle of gender. Instead of assuming the responsibility for the well-being of the race he turned that function over to the government while he pursued material gain. A patriarchal structure never developed in Europe; the non-sustainable nature of the nuclear family established in its place requires government intervention to maintain it. Old age homes, orphanages, and prisons are the norm of Western societies. The institutionalized care of human beings has become a major European export.

By every mental, physical, and spiritual measure Western society is crumbling. In order for people to improve their lives under these conditions it is necessary to live outside of the system. The mission statement of Men's Action to Rebuild Society calls for the fostering of a more natural way of life. The establishment of a patriarchal societal structure serves as the primary step in that direction.

THE GENDER SIGNIFICANCE OF THE AQUARIAN AGE

Our society personalizes and materializes every bit of knowledge and truth that it comes across. While this is the natural characteristic of the feminine principle it nevertheless, disturbs me that nothing can be seen through masculine eyes or even considered with a gender perspective. While ruminating over this matter I realized that even the transition into the Age of Aquarius has been heralded through feminine, passive, and materialistic thought—It's something wonderful that will just happen to make the world "nice" while we just sit on our behinds and wait for the event.

The transit from the Piscean age into the Aquarian age is a gender dynamic that I have not seen anyone address, most likely because we live in society that has little understanding of the universal principle of gender. While I have only a layman's understanding of astrology and astronomy, I do understand the universal dynamics of gender and believe I would be remiss to our organization and the propagation of its message if I did not relate this transition from one age to the other, to gender.

The Aquarian Age has received much attention during the past 50 years since its public introduction into society by the hippie movement. The song The Dawning of the Age of Aquarius, theater presentations, magazine articles, art exhibits, and books have trumpeted the coming of the new age. Astrology, numerology, clairvoyance, mysticism, the occult, holistic healing, shamanism, and palmistry have the designation of "new age" subjects, even though they have existed since the beginning of our present age.

Some say the Aquarian Age began 50 years ago, some say it began in 2000, and others say it is 500 years away. Regardless of whose calculations are correct, the slipping of the Solar system backward into another zodiacal sign occurs about every 2,300 years and is called its precession. We have been in the Piscean Age for the past 2,000 years or so and are now experiencing the ending of that age in preparation for the "new age' of Aquarius.

The sign Pisces is a feminine, mutable, water sign. The code name given to Pisces by astrologers is "I believe." The Christian era, which began at the beginning of the Piscean age emphasized water (baptisms, fish, and fisherman), and salvation through faith (belief). The Piscean personality being mutable and feminine receptive, requires authority figures for it to function properly. All aspects of Western society are built upon authority whether, religious, governmental, legal, medical, educational or scientific.

The Western influenced Piscean society has no ability to resolve problems or issues. Feminine characteristics respond to conditions, adapt to them, and express their feelings about them; they do not change them. Hence the writers and speakers of today elaborate on the wrongs of society but are unable to articulate what needs to be done to bring about positive change. All of the political parties and activists are impotent. They have no ideas. The self-emasculated Western male's ability to lead has been destroyed. The change that people desire will not take place.

Without masculine direction chaos develops. Consequently the authority structure of Western society keeps enacting more laws and becomes increasingly involved in the private lives of people in order to maintain some semblance of order. Yet with each usurpation of individual control a new need develops for still more authoritative influence. Western society is close to the point of complete tyranny; which will be the final form of government before its total and complete demise. Fortunately the Age of Aquarius is dawning.

The sign Aquarius is a masculine, fixed, air sign. The code name given to Aquarius is "I know." The Aquarian personality only accepts what it can understand or experience. As an air sign it emphasizes the intellect, and as a fixed sign it is goal oriented and exhibits the constancy and reliability necessary to bring about these goals. The Aquarian psyche has a universal outlook and will work for the betterment of humankind. As a masculine sign it has the ability to deal with the unseen and the virility to initiate change.

The Western concept of the dawning of the Age of Aquarius envisions an increase of the feminine influence, the feeling of the pulse of mother earth, and the equality of the sexes (with the women being just a little more equal). Instead, I believe the present feminine oriented Western society will collapse and/or be blown away and then be replaced by a masculine oriented rebuilding that will strive for the betterment of humankind and a return of an understanding and implementation of the natural patriarchal structure.

Western society will not be "fixed," it will come down. Pisces resides in the 12th house, which is also known as the house of self-undoing, and Western man has with his own hands cast himself into destruction. Society will be rebuilt with Aquarian overtones.

This is my brief gender interpretation of the transition that is taking place as we leave Pisces and enter Aquarius. Most likely you will not readily find a corroborating viewpoint for reasons that I have mentioned.

THE HUMAN JACKASS ERA

American history, like all history is punctuated with eras signifying personalities, ideas, and practices that define the times. It has had the Reagan era and the Johnson era, the Clipper ship era and the Railroad era. It also had the era of the Horse Soldier, which made possible the conquest of the plains Indians and added to the lore of their famous Chiefs.

America and the entire Western world has moved into the Human Jackass era.

A jackass is usually kept in a barn overnight and in the morning the farmer brings it hay to eat before commencing the day's work. The farmer provides the on the job training for the jackass, which tends to learn rapidly and becomes a reliable worker, often without use of a tether. If the animal is injured or sick the farmer provides for its healing and when necessary will call upon the services of a veterinary. The jackass normally does not mate unless the farmer brings in a she-ass for that purpose. After mating the farmer decides what to do with the offspring. Life is fairly simple for the jackass; the farmer provides for its every need—as long as it works.

Western man now lives the life of a jackass. He has been stripped of all his natural male authority, which has resulted in the demise of the family and reduced his function to that of doing the work that some higher authority dictates. He receives medical care in a form and degree dictated by the government. The state provides for his education, which in one way or another consists of vocational training. When he does mate the state takes over the care and training of his child.

Western woman, once revered for being a devoted wife and nurturing mother has been stripped of her prestigious position due to the elimination of the family. What is she to do? The government has told her not to worry that she has a right to do the same work as the men, and as verification of this right she will be given diplomas, licenses, and college degrees all of which serve as she-ass certificates that will enable her to do work similar to that of men. She-asses now toil on the treadmill of production as equals to the jackasses.

There are many people who like this beast of burden arrangement. They clamor for universal healthcare so that they will no longer have the responsibility of taking care of their bodies; they clamor for aid to education so that they will not have to prepare for their future; and they clamor for surveillance cameras and security police in an attempt to develop the safe environment once provided by men.

Many she-asses are content to have the state feed, educate, and provide healthcare for their offspring, which gives it the opportunity to indoctrinate them and thus proliferate the number of people who accept the jackass way of living. By the time state-reared children reach adulthood they look upon government as the source of all things and are quite content to become beasts of burden and let the state provide for all their necessities.

The human jackasses, not having families to care for them must rely upon government care when they can no longer pull their loads. Government provides retirement plans of various sorts and social security to help the aged, but eventually the effectiveness of these plans diminishes due to the erosive effect of government created inflation. No longer able to sustain themselves the aged are put in government housing and fed canned and packaged meals that have very little life-force in them. The aged develop reactions to the poor quality of nutrition, which the government treats with medicinal drugs. Eventually these aged bodies can no longer cope with the toxemia of poor diet and pharmaceutical products and they develop "incurable" illnesses attributed to their age. Since families have been eliminated, the government makes the life and death decisions for the aged.

The jackass era has no culture for it was stripped of any traditional practices relating to the propagation and preservation of the species as part of its training for the workplace. Entertainment has become strictly sensation in one form of licentiousness and debauchery or another. These escape mechanisms provide only short-term relief from the boredom and un-natural life of being a jackass or she-ass.

All is not well in the human jackass community. Ten percent of the she-asses in America and 15% of those in Europe suffer from debilitating depression They are slowly learning that the government is not a good substitute man and/or husband. Mental illness has become the number one health issue in the English-speaking world. Some jackasses have begun to chafe in their stalls as they become increasingly dissatisfied with their un-natural environment.

Eventually the more enlightened of the jackasses will look into the mirror and see that they created their environment by buying into the materialism of Western thought. The indwelling spirit causes a realization in each one that life consists of more than the acquisition of things. It awakens him to the higher purpose of spiritual growth and the importance of the propagation and preservation of the species. He comes to the realization that to consider those with testis and those with teats as being the same reflects asinine thinking and it was only natural that he began to live the life of the jackass as a consequence.

Realizing his responsibility for his present situation he becomes aware that he also has the power to alter it by changing his thinking and having the courage to live out those changes.

Some believe a new spiritual era will be ushered in by divine intervention. Such thought—the expectation that some other force will take care of things—is part of the passive feminine thinking that led to the jackass era and the institutionalization of human beings. The universe is designed to operate in a certain manner. What is in harmony with the universe endures, what isn't doesn't. The society created by Western thought is an aberration and is imploding. When it collapses the "opportunity" will be there for men of courage and spiritual insight to start rebuilding a more natural society.

Change begins with the individual and societal change requires the cooperation of others. Like-minded men working together will break free of jackass thinking. They will unhook women from the treadmill of production and return them to an environment focused on the family that will permit them to fulfill their natural purpose of bringing life into the world and nurturing it. These few men will be known as the rebuilders and in the words of Isaiah they will rebuild on the old foundations. They will realize that the propagation and preservation of the species is the highest material calling on the journey of spiritual unfoldment and that patriarchy provides the structure that enables the establishment and support of the extended family.

The time is at hand for those men and women of insight and courage to rid themselves of the jackass mentality, get off the treadmill of production, and start living the lives that they were ordained to live.

It is time to spread the word of another way of life, which can be done by sending this essay to newspapers, magazines and various organizations that have an interest in family and personal accountability. You can invest in the various information materials offered by Men's Action and distribute them to friends, family, and associates. You can subscribe to the Men's Action website and to its videos on youtube. No one is going to make the change for you. You and like-minded people working together can provide the change that all of humankind is thirsting for. Now is the time to start.

THE LEARNING FAILURE IN THE SCHOOLS

Wisdom is not finally tested in schools
Wisdom cannot be passed from one having it to another not having it
Wisdom is of the soul, is not susceptible to proof, is its own proof

Walt Whitman

For millennia childhood education consisted of learning how to become a well functioning member of the family and the tribe by doing one's share of the necessary work and looking after its members in a caring and considerate manner. With the destruction of the family children no longer receive training in these values and thus arrive at school with a relatively blank slate of functional ability and moral standards. The educational system has not made any attempt to compensate for this loss and has set as its goal the training of children to be production workers. Teachers do not use the term "training to be production workers", instead they substitute the term, "You must have an education so that you can get a good job and make money and become somebody"; the inference being that children are nobodies, and through a grossly materialistic approach they will be made into somebody's.

The feminine materialistic approach of the educational system abounds in malpractice; however, nothing has had a more devastating effect on the development of children than the practice of coed education, especially during the post maturation and puberty ages. Of the many gender characteristics, it is the difference in thinking that most significantly affects the ability of instruction to be compatible with one sex or the other. Since females think visually and males think both visually and conceptually, in order to accommodate both sexes in the one classroom, education has been geared to visual thinking, an unnatural environment for males.

The most common phrases of females in the classroom, "I don't see" and "Not always" in response to what they don't understand, constantly nettles the males. The self-evident first phrase requires the breaking down of the subject matter into seeable bits. The "not always" phrase serves as a frequent reminder that they do not understand the concept of generalities (remember the feminine principle thinks personally). A steady diet of communication and inter-action in a manner not natural for males causes them to want to leave the academic environment. At the college level men have free choice as to whether to continue their education, and many have chosen not to continue. Male college enrollment at this time has dropped to 35% and is expected to drop to 25% within three years. High school males do not have the choice of dropping out but truancy rates and disciplinary problems have risen. At the grade school level boys do not even have the luxury of truancy to escape from their environment and are drugged so that they can better endure it. Close to one million boys receive the drug Ritalin on a regular basis in order to sedate them, their big crime being that they don't behave like girls.

One grade school boy who didn't behave like a girl was Thomas Alva Edison. Al did not take well to rote learning and at the age of eight his teacher considered him to be a little mixed-up. His mother took him out of the school that found him defective and read classics to him and put him into another school, which he left for good at the age of 13. Thomas Edison was not only an inventive genius but also an entrepreneurial genius; companies such as General Electric and Con Edison owe their existence to him. If Edison had attended school in today's environment he would have been given Ritalin. How many Thomas Alva Edison's has the educational system destroyed?

In addition to the un-natural academic environment thrust upon boys, they must also deal with an athletic environment that forces them to compete with girls. Since the beginning of time males have been trained to protect women, for they brought life into the world and nurtured it; now males are forced to compete physically with those whom they were designed to protect. This policy is beyond ignorance; it is madness.

This madness is compounded by the intense sexual tension of the coed academic environment. Gender exists for the purpose of reproduction, which is done by mating, and most teenagers readily arrive at an understanding of how this mating takes place. In a society that raised girls to become wives and mothers and boys to become husbands and fathers, care was taken to prevent premarital sex, or at the least delay it. Girls were protected from the sexual advances of boys by parental supervision, the chaperoning of social events, and the minimizing of contact between the sexes by having single sex schools. However, with materialism being the sole motivation of the liberal wimp and his feminist camp followers, and the equalization of the sexes a part of their dogma, the only preventive procedures taken were to inhibit the natural assertiveness of males by enacting sexual harassment codes. Beyond that, the only instruction for youth were pregnancy preventive measures, and if they didn't work, the availability of abortions without parental consent. This is evil.

Children denied a natural home life and thrust into an unnatural learning environment have little potential to absorb academic material; however, the Western culture relying on the three material powers of money, information, and police authority, attempts to force-feed information into our children. Yet Johnny can't read. More money is budgeted for higher salaries for teachers, more modern classroom equipment, and smaller class sizes. Yet Mary can't count.

Talking about an issue and throwing money at it, doesn't work with education. The Washington, DC school system serves as an example. In the early 1990's it had the highest per pupil expenditure for education, the lowest pupil/teacher ratio in the nation, the sixth highest paid teachers in the nation, and a per capita aid for education that surpasses the national average by 250%. The results of these expenditures were that on time graduation rate ranked 49th in the nation, math proficiency for 2nd and 8th graders ranked last in the nation, reading proficiency for 4th graders ranked last in the nation. More money spent on education had no noticeable effect on academic performance in DC.

There are other factors that affect academic performance, such as the child's environment. DC has the highest national rates of crime, assault, aggravated assault, larceny theft, vehicle theft, murder, and robbery. Also, in 1990 the percentage of one-parent families was the highest in the nation at 54.7, and the percentage of births to unwed mothers was the highest in the nation at 64.9. Both of these statistics have gotten worse.

The DC school system exemplifies what happens to student performance when fathers are absent from the home. Conversely, when children have fathers at home the school drop out rate is cut in half, the teenage pregnancy rate is cut in half, truancy decreases, academic performance rises, and juvenile delinquency decreases. Money is an ineffective substitute for fatherhood. The DC school system could not fool the children.

The information on the DC school system was taken from the text *"DC By The Numbers: a state of failure"* by Thomas H. Edmonds, but other sources come to the same conclusion that Mr. Edmonds did. Author Susan Meyer conducted research with the intent of proving that children from poorer environments would do better in school if the expenditures for education were increased, and her findings showed this premise to be wrong. In her book *"What Money Can't Buy"* she indicates that the factors that resulted in positive academic performance were the home environment, the value structures that it promoted, and the motivation of the children. Strong and loving families provide the environment and motivation for children to learn well and behave responsibly.

This penchant for relying on absolute materialism to educate our children has caused me to reflect upon my youth when I knew an immigrant couple that settled in the mountains of Pennsylvania and started a small farm and a large family. They had 11 children and were so poor that the boys and girls alike wore dresses because they were easier to make and alter. The children, six girls and five boys, were educated in a one-room school that had only one teacher. They received no special educational training from their parents at home because their parents were not well educated. As adults all of these children were employed, some in their own businesses, and one became a nurse. They were all literate and functioned well socially.

What these 11 children had that the majority of children do not have was a mother and father at home with a sense of ethics and morality that became inherent in their lives. They saw their mother and father work, an activity that illustrated a relevancy between what they were taught in school and what was necessary to accomplish the tasks of their parents. Those children knew how to milk cows, churn butter, raise vegetables, cut wood, hunt, fish, water and groom the horse, cook, bake, clean and do a host of other activities that required judgment, decision making, and the use of what they learned in school.

Compare the environment of those farm children to the average child today, who only rarely has two parents at home, and even when they do, finds them functioning as production workers rather than as men, women, mothers, and fathers. Well functioning school children come from well functioning homes. Money cannot buy that and legislation cannot enact it.

I have not gone into the specifics of school instruction because at this point it should be clear that the damage done to our children by the destruction of the family coupled with the un-natural school environment makes them incapable of absorbing a meaningful education. For those who desire more specifics of classroom instruction John Taylor Gatto's book, *Dumbing Us Down* might be considered a bible of the failure of compulsory schooling. He and his associates such as Thomas Moore and David Alpert, who respectively wrote the forward and introduction to the second edition, recognize the necessity of mixing young and old people together, the importance of family, and the dehumanizing effect of our compulsory schooling system. Mr. Gatto not only points out what is wrong with compulsory schooling, but also offers many alternatives. His promotion of home schooling has motivated many families to give it a try and with various levels of success. However, how can home schooling be conducted when there is no home? With only 18% of children 18 years old living in a home with both their natural parents that means 82% of children come from a broken or altered home; an environment not conducive to home schooling. Of the 18% of households that still remain in tact, many have both parents working outside of the home, a situation not conducive to providing home schooling. Mr. Gatto and all his competent and well meaning associates fail to get at the heart of the matter, and that is that family can only exist without government interference when it is an extended family and that can only occur in a patriarchal polygamous structure.

THE TYRANT MOTHER

The tyrant mother—an issue that reached major proportion in the 20th century—has affected human relationships throughout the Westernized world. The tyrant mother has existed since the earliest of times, however, the proliferation of this condition has resulted in a modern phenomenon that needs to be addressed and corrected. And soon!

The dictionary defines tyrant as an absolute ruler unrestrained by law or constitution. The tyrant mother has absolute authority in a home devoid of the masculine influence. Put another way, an ordinary woman who raises children without the presence of an effective masculine figure becomes a tyrant mother. The word effective is stressed, for a weak or ineffective man is worse than no man at all. Many women have come to this realization and have opted to raise children without living with their fathers. Unfortunately, they most likely have become tyrant mothers.

It is easy to spot the tyrant mother; she will berate her children in public, smack them around, and scream at them, or she might have them under such control that they won't make a move without her permission. Tyrant mothers have been depicted in African-American films such as; Boyz N The Hood, Mo Betta Blues, and Do the Right Thing. In Boyz, Ice Cube's mother kicks his butt and yells at him frequently. Was she a bad woman? No. She had made an outdoor barbecue for the teenagers in the neighborhood. She did what most women do – nurture the race. But she wasn't designed to discipline two teenage boys in a tough neighborhood; that's what men are supposed to do. In Mo Betta Blues, Denzel Washington's father was a nice guy, a dandy. He was also ineffective. Mom had to be the disciplinarian. In Do the Right Thing the tyrant mother smacked her son all over the street. As these films accurately depicted, children of the tyrant mother rarely do well.

In his novel, *Portnoy's Complaint*, author Philip Roth described the negative effects of being raised by a tyrant mother, but the critics and society didn't recognize it as such. They thought the problems resulted from the influence of dominant Jewish mothers. The issue of the tyrant mother is no more Jewish than it is African-American, or Hispanic, or Asian. Tyrant mothers come in all flavors, religions and ethnicities. They have one thing in common—the absence of a competent masculine presence in the home.

One of the many meanings of the story of Adam and Eve is that Eve paid the price of Adam not fulfilling his responsibilities. Women always pay the price when men do not fulfill their responsibilities. The failure of Western man to recognize his function as the head of the family has resulted in the frustration of mothers, the neglect and abuse of children, and the dissolution of the family.

The first massive wave of tyrant mothers came into being early last century, when men working long hours and gone from the home 12 to 14 hours a day six days a week, didn't live at home; they just ate and slept there. Mom did everything else. She also became a tyrant. However, the major development of the tyrant mother was in the last third of the century, when the government's "Great Society" legislation emasculated the male by removing him outright from the home or leaving him with no authority in it. The results have been devastating to society.

THE YOKE OF SLAVERY DESCENDS UPON US

In my article Withdrawal Symptoms I wrote, "A conspiracy film that I saw recently contained the statement that the way to enslave people is to get them to want to be enslaved." As luck would have it a current example of this viewpoint came from the Mayor of New York during his visit to London in which he admired the use of surveillance cameras all over the city. He stated, "We live in a dangerous world, and people want to have security cameras."

Let's focus on the significance of this statement. We have the government, which has taken over the responsibility of the male to care for the safety of women, children, and the tribe, telling the people what is good for their protection and then getting them to want it. And for doubters such as me he states, "It's ridiculous, people who object to the use of technology." "The MTA just has to get us this technology. Americans are too exposed and there are some people who don't like cameras, but the alternatives could be worse." That's it Mayor Bloomberg, scare the people into wanting to be enslaved.

The MTA (Metropolitan Transit Authority) will indeed get us this technology and we will become grateful for another arm of government enslaving us even further.

Another aspect of this onrush to our enslavement that requires our awareness is that we have not been asked, but have been told what we want and what will be done to fulfill that want. Very few laws come about as a result of asking the voter anything. I wasn't asked if I wanted daylight savings time extended, or if 2.6 million condoms should be distributed to our school children, or about any other law that I have to obey. Government enacts laws as they see fit and only the power of lobbyists on one side or the other can alter these laws before enactment. The removal of manly power results in his enslavement by the state. Women do not realize this truism. Their focus is on security, and since most of them no longer have a man in the house, whether a father or a husband, they welcome any security system offered by the government, the very instrument that emasculated the male and made the female clamor for security.

You don't have much time in which to act my brothers. The complete control of your life by the government is close at hand. The yoke of slavery is descending upon you. Once it happens, the ability to break out will be extremely difficult, if not impossible. Do you intend to sit on the fence until it happens, or are you ready to stand up and do something about it now?

You know where we are.

THE WESTERN PENCHANT FOR THE LAW

Websites all over the internet contain multitude blogs that refer to the failure of our laws and how they need modification or supplementing. Western man believes that good laws make good people–a belief system that has no basis in reality. Good people need few laws; that has a basis in reality. A tribal way of life devoted to the propagation and preservation of the species develops good people; it's called patriarchy. That has plenty of reality–it can be seen everywhere.

The constant references to the law call for a way of life hoped for or that once was but has not been seen. It is talked about or imagined. Who saw the laws the way he would like to see them? When and where did they exist? Even Madison and Jefferson wrote comments about the ineffectiveness of the Constitution. Western man never had a society in which the laws promoted the family and the tribe. That's one of the reasons all his societies collapsed.

Those who grew up in a rural environment might say they can remember when the laws worked, but agrarian people live essentially without the law. The closer people get to working the land the more natural is their division of labor and practice of gender. Men clear the forests and plow the land and keep the homestead safe. Women churn the butter, cook the food, make the clothes, and do other forms of work that also enable them to watch the children. Agrarian men do what patriarchy does; provide the means and the environment for women to bring life into this world and nurture it. To those who have lived in it or even seen it, consider yourselves lucky.

Patriarchy is something I have seen in cultures allover the world. I have seen patriarchy at work among Greeks (they got it from the Ottoman Turks), Italians (they got it from the Carthaginians), Koreans, Kenyans, and Pakistanis, just to name a few. The patriarchal-polygamous structure exists everywhere. The relation of the planets to the Sun and the electrons to the nucleus of the atom are patriarchal-polygamous, as are the relationships of the bull to the herd, the rooster to the flock, and the lion to the pride. The entire universe is patriarchal-polygamous. Only Western man thinks differently.

In my information booklet, *20th Century Decadence: 2000 Years of Western Materialism,* I indicate the following*: A society that has no spiritual basis, that relies on materialism, and that does not practice some form of tribalism, will have no structure to provide it with cohesion. In its effort to substitute for the lack of the masculine influence, the European culture relies upon the law.*

From the earliest European Greeks there existed a great dependency on the law. Draco instituted a system of law based on an aristocracy (the feminine form of government.) Solon developed a system of law in which eligibility for political power was based on wealth (a feminine value.) From Draco, to Solon, to Justinian, to Napoleon, the Europeans prided themselves on their legal institutions. After a 2,000-year history based on law, when the Europeans came to America the Indians said of them, "white man speak with forked tongue." Indeed he did, for the law is but a poor substitute for the masculine influence. The manly way is "let your yea be yea, and let your nay be nay." The law is for those who can't live by that code.

Western man has legislated away all of his God-given authority and turned it over to the government; now he is crying about it and believes that if he cries hard enough and long enough about it the government will bring about change. That's what women do when they want a man to change something; it is not what real men do; they know they are the ones who bring change about. That was one of the allegorical meanings of the story of Adam in the Garden of Eden. Adam represented the masculine principle. He was given all authority. There were no governments, no laws, and no rights. There were obligations, consideration, and responsibilities.

The laws of this land will never change for the better–they never have. Only men can change this nation for the better. They will have to stand up individually and work collectively for change. Or they can do nothing and drown in a sea of degeneracy. There are but two options and one choice. What is yours?

WAITING FOR THE JAILER

The American incarceration rate has reached a new high. Now more than one man in every hundred is in prison.

For those who have checked my streaming video you will have noticed that I start off every video pointing out the high incarceration rate of men. I started filming in 1993 and at that time the number of men in prison was just over one million. That was a 400% increase over 1976. We now have 2.4 million men in prison.

And what are we doing about it? Are we just waiting at home for the KGB to knock on our door and take us away? Perhaps not. Instead we might be so grateful that we are living in the land of the free—the land that has the highest incarceration rate in the world and more men in prison than China and India combined—that we do not even notice this infringement on our liberties.

We have passed the point of living like the people in George Orwell's *1984*; we have moved to the level of *Huxley's Brave New World*—we love our suppressors. Anyone who says anything negative about our suppressors is un-American.

For those who believe in working within the system, putting people in jail is what the system does. It has been doing this on a consistent basis ever since the Great Society legislation of the 1960's. It is a way of life in America. And this high rate of incarceration is occurring in good economic times. What will happen when the economy finally collapses and unemployment becomes rampant?

Apparently we'll just wait and see. The lack of response to my last Message From Elder George would indicate that to be the prevailing viewpoint. We'll just wait. Maybe it'll go away. Maybe they won't get my family or me. If it hasn't gone away in the last 50 years, but instead has gotten worse, why would anyone expect things to get better?

We live in a society that has opted for the rule of government instead of the rule of men. The way governments maintain their rule is to put men in prison. They do this by enacting increasingly onerous laws and then arresting those who transgress those laws.

If we don't want men to be put in jail in increasing numbers we have to start ruling ourselves. That means working outside of the system. There is only one organization that offers a solution outside of the system. It's called Men's Action to Rebuild Society. It's not an organization that is going to make change for you; it's an organization that will give you the opportunity to help change your circumstances with those who believe as you do.

Men's Action requires more than your support; it requires your involvement. It means getting rid of the expectation that an outside force will better your life. It means that you are responsible for your life and the greater community. This is not done through government but through patriarchy and the care of family. It means fulfilling your manly responsibilities.

There are no dues at men's Action, but it is expected that you will promote the message that you believe in, which is done by distributing our book and booklets. I suggested a starter supply of two books and two sets of booklets, one for you and one for an interested person.

You can invest in change or wait for the knock at the door

WESTERN RELIGIOUSNESS & THE NEW DECADE

Religions are feminine; they materialize and personalize impersonal and unseen spiritual truths. In theory they represent the collective understanding of the populace regarding the teachings of prophets contained in various texts; in actuality they represent the belief system of those in charge of the religion. The less understanding a group has of matters spiritual, the greater the need for a belief system and the authority necessary to enforce that belief. Western materialistic thought has little understanding of matters spiritual (the unseen); therefore, its religious structure relies heavily upon authority, law, and punishment.

Western thought develops religiousness in all its institutions. The medical field has little understanding of health; therefore, it developed a religious structure to sustain itself and enforce its beliefs. What the AMA refers to as a protocol is a liturgy prepared by drug companies in collusion with attorneys and enforced by the high priests of a medical oligarchy who demand absolute adherence to the liturgy with the threat of hell and eternal damnation to the physician who disobeys. That means the loss of his license.

The government has little understanding of unseen ethics; therefore it has no basis for developing moral behavior. It focuses on legal and illegal, for it does not know right from wrong. It has created a huge religious institution called the American Bar Association, which has authority, laws, and punishment. Those who do not meet the criteria of the high priests are disbarred.

The educational system has little understanding of knowledge, which is unseen; therefore, it focuses on information, which is seen and develops religiously oriented institutional structures based on authority, law, and punishment. It will only consider new information if it is based upon the approved liturgy (materialistic scientific approach) for developing and presenting this information. Those within the institutions who do not conduct themselves in accordance with the accepted academic liturgy face expulsion, or if tenured, ostracization. Those who attend these institutions will be fed a diet of rote learning and examined on the retention of the information fed to them.

Western thought has little understanding of the unseen forces of nature; therefore, it ravages, pollutes, and distorts the environment. In order to administer the use of natural resources (that's the name it gives to nature) it has made of it a commodity controlled by international finance. These moneyed interests not only bought Manhattan from the Indians, they bought the world; they own it and decide who lives where and how. They have created an international economic cartel that has its authority, laws, and punishment. They enforce a belief structure predicated on the power of money.

Every activity that Western thought engages in is based upon what it sees and believes; not upon what it doesn't see or understand. The entire world of Western thought is a vast empire of religiously structured institutions based upon gross materialism, which in one way or another acts to enslave the populace.

The major manifestation of Western thought is the lack of understanding of the purpose of human existence. It does not "see" that we are here to propagate and preserve the species on our journey of spiritual growth. Therefore, it does not understand the purpose of gender and the patriarchal structure necessary to attain that objective. Instead Western thought believes in the finiteness of the material world and the unlimited opportunity to indulge oneself in it. This thinking has created a religious materialistic governing structure with a multiplicity of institutions with authority, laws, and punishment designed to propagate and enforce its belief.

These institutions by nature attempt to expand their power. Ecumenism attempts to create a one-world religious order. National governments attempt to create a one-world social order. Financial institutions attempt to create a one-world financial order. The collective goal is to have an institution to govern every facet of life on an international basis. Western thought considers that to be absolute fairness; it does not realize that it is absolute subjugation of the people. It eliminates the necessity for discernment and replaces it with the necessity of absolute obedience.

Western thought has enslaved its society through the religiousness of its institutions.

Various men's groups around the world (the Western world) are attempting to fight the punitive laws directed against men and family. They attempt to do it through the very institutions that emasculated them. These men's rights groups are now fighting for parity with women. Imagine that; the Adam principle of the universe is begging the institutions that it created to treat them the same as women. This thinking leads to the impotence of society. "Lord, forgive them for they know not what they do."

However, all is not doom and gloom. The universe is always in divine order. What is natural endures; what isn't doesn't. The society and institutions created by Western thought are in their death throes. We are witnessing its demise in real time. As it collapses those that remain will be forced to see the thing behind the thing—the spirit behind the material world. Society will recognize its purpose to propagate and preserve the species while on its journey of spiritual development. Men will realize their obligation to provide the environment and means for women to bring forth life and nurture it. Instead of focusing on religion, humankind will focus on spiritual understanding. Instead of focusing on the law it will realize its obligation, responsibilities, and considerations toward its brothers and sisters, fathers and mothers, and sons and daughters.

Prior to the collapse and the enforced change that it will produce, a small and virile minority with more enlightened views will begin to make its presence felt. Those minorities of people are with us now, and their numbers are growing. They have been reborn; their thinking has changed. They consist of women happy to be women, and men cognizant of their manliness. Working together these men and women realize their responsibilities, obligations, and considerations to the human race. They rear their own children rather than having the state do it. They have ethics that govern their moral conduct. They recognize the impermanence of the material world and have an understanding of matters spiritual. They have an appreciation of nature and the interdependence of all life. These people will serve as the nucleus upon which the new era will be built. The small and growing minority with us that will meet that challenge is fulfilling the prophecy contained in Isaiah, which states that they will rebuild on the old foundations and will be known as the rebuilders.

As the year 2009 drew to a close we welcomed not only a new year but also a new decade, one that will usher in a period of change perhaps unparalleled in human history. We will see the institutionalization of human beings reach its peak and then crack and implode as the impotence it created gives way to the growing virile minority that will be the cornerstone of a new era.

MEN'S ACTION, INC. MISSION STATEMENT

To foster a natural way of life for humankind.

Objectives:

I. Rebuild the family
II. Establish a community based tribal structure
III. Promote a spiritual and material balance
IV. Foster respect for the environment and a life style that results in health and well being

Objective I: Rebuild the family.

Strategy:

A. Create an awareness that men make families and, that patriarchy produces extended family
B. Conduct male leadership training seminars and programs with emphasis on family and community responsibilities
C. Develop a spirit of camaraderie among men
D. Conduct single sex youth education programs and develop in each sex their unique ability and responsibility to the family

Objective II: Establish a community-based tribal structure.

Strategy:

A. Have street representatives address the personal and community concerns of the residents.
B. Work with residents to inculcate in all people, especially youth, respect for others and their property.

C. Cultivate a respect for elders and involve elders in the counseling of members of the community

Objective III: Promote a spiritual and material balance in social values and activities.

Strategy:

A. Cultivate the realization that the purpose of the economy is to provide for the nurturing of the race, that it is not an end in itself
B. Illustrate that all material things become obsolete, and that it is the unseen values that endure
C. Promote a respect for other cultures, ethnicities, and religions

Objective IV: Foster respect for the environment and a life style that results in health and well being.

Strategy:

A. Promote a toxic free environment
B. Support organic farming
C. Promote vegetarianism
D. Promote holistic health

www.ingramcontent.com/pod-product-compliance
Ingram Content Group UK Ltd.
Pitfield, Milton Keynes, MK11 3LW, UK
UKHW041826200726
13854UKWH00002BA/605